# CONE SHAPE AND COLOR VISION:
## UNIFICATION OF STRUCTURE AND PERCEPTION

John A. Medeiros

John A. Medeiros

**Cone Shape and Color Vision:**

**Unification of Structure and Perception**

First time or interested authors, contact Fifth Estate Publishers, Post Office Box 116, Blountsville, AL 35031.

First Printing May 2006

Cover Design by Three Fold Design

Printed on acid-free paper

Library of Congress Control No: 2006927333

ISBN: 1933580224

Fifth Estate 2006

# CONE SHAPE AND COLOR VISION:
## UNIFICATION OF STRUCTURE AND PERCEPTION

**Table of Contents**

# List of Figures

# *Preface*

Pick up any textbook or article on vision from the current literature and chances are you will see the statement: "Human color vision is based on the existence of three types of cones, a red, a green, and a blue sensitive type", or very similar such wording.   There is no discussion, no question, just this simple statement of "fact".   With such an oft repeated and widely entrenched description of human color vision, there is surely no question that it is right.  Or is there?

This volume represents many years of effort on the part of the author as well as of a number of collaborators to try to better understand how human color vision works.  Some of our work has been published in standard journals, but the key conclusions we have reached about color vision have run up against a very entrenched resistance and have not appeared in a unified form elsewhere.

I should first say a bit about how I got into the effort that has resulted in this volume.   I did my graduate work in atomic physics at the University of Massachusetts and on graduating felt like I wanted to do something else.   So in the early 1970's I took a position as a postdoctoral fellow at the University of Western Ontario where I could use my physics background in a project to study laser damage mechanisms in the retina.  Since I knew very little about the eye, I began a program of reading everything I could find about the eye and its workings.  As a physicist I gained an enormous appreciation of just what an engineering marvel of design was the human eye.  As I read each little bit, things would occur to me like 'I wonder if anyone has thought of that before…' and as I read on I would find that, sure enough, someone had indeed thought of 'that', whatever it was, many years earlier.

This kept happening as I learned more and more about the eye and vision until I started looking at the details of size and structure of the retinal cones.  As a physicist I wondered if anyone had thought about what should be an obvious connection between the cone form and its function in terms of a simple wavelength dispersion mechanism that should be present in a properly sized and tapered waveguide. I thought then that 'I'll just read some more and find that this has already been thought of long before and, since it's not discussed anywhere in the current literature, it must have been discarded for some reason'. Well, I read on and was astonished to see that this had evidently not been mentioned previously and that nowhere was there any discussion or thought about the connection between cone shape and cone function. I then figured that I had better look at this concept in more detail and see what must be wrong about it. I did that, but the closer I looked, the more I became convinced that it must be right.

With a good deal of anticipation, I then tried to put the arguments for this concept connecting shape and function of the cones together and submit this simple and apparently novel idea for publication in the most prestigious scientific journal I could think of. I had naively expected to receive a welcoming reception for a potentially important new idea. Instead, I received the response, paraphrasing one reviewer, "What is needed now is not a new model; we know how color vision works, we just need more good experimental work confirming what we know to be right". The establishment had spoken! Over the years, I kept expecting that someone would eventually think about this concept and finally put it in the literature. Astonishingly, that has not happened so it occurred to me that I had better do my best to put all the pieces together and get this concept and its consequences published in one way or another. The result of that effort is the volume you have in your hands.

This book will not be immediately welcomed by researchers in the color vision field since it proposes a model that goes against the grain of a basic tenet of current color vision science: that human color vision is based on the presence of three cone types in the retina. In common parlance the three cone types are generally termed the Long, Middle, and Short wavelength sensitive cones (L, M, and S cones) since their putative sensitivities don't match up very well with the red, green, and blue color namesakes.

I will address those issues that are commonly invoked to support the three-cone model of color vision and suggest that the actual evidence leaves things somewhat wanting. Indeed, there exists some key, but widely ignored evidence, that flatly contradicts the basic tenets of the three-cone model. In the following discussions it will also be important to keep in mind a critical distinction: three cone types and three pigments are not necessarily synonymous concepts. Some of the key data invoked to support the three-cone model comes from molecular genetic studies that have identified genes that code for a number of cone photopigments. However, it is an entirely separate matter to establish that these photopigments are each segregated within their own unique cone type and operate in the way classically envisioned for the three-cone model of human color vision. Multiple pigments do not necessarily mean multiple cone types. There could, for example, be some mechanism that uses multiple pigments in a supplementary role to enhance or optimize some other means of providing the basic discrimination of color vision.

I suggest exactly that and describe here a simple theory that makes use of spectral dispersion by tapered optical fibers of the appropriate size and shape to encode color information along the length of a cone. This dispersion mechanism is demonstrably a direct consequence of the basic physics of optical waveguides with the appropriate optical dimensions. We will see that the cone photoreceptors of the human retina are correctly dimensioned to exhibit this effect and that there is a simple and natural way in which the color information available through this cone spectrometer action can be neurologically coded and used by the retina and brain.

I will further show that this mechanism explains, in a very direct way, a number of hitherto very puzzling aspects of human color vision (such as subjective color effects and the perceptual similarity of violet and purple) and finally provides an answer as to why, unlike the cylindrical rod receptors mediating monochromatic night vision, the color receptors of the eye are tapered, that is, the cones are conical. Indeed, hitherto there has been no explanation for one of the most basic, distinguishing characteristics of the color receptors – that they are in fact cones. It would seem rather preposterous for nature to universally employ the cone shape for color receptors simply to enable us to distinguish them from the monochromatic rod receptors!

This volume describes some original research that provides significant support for the proposed model. One part of this research is the direct observation of this color dispersion effect occurring in tapered optical fibers in precisely the way predicted for the retinal cones (photographs shown on the front cover). A second piece of original research is the direct observation of the separate perception by the rods and cones of the retina. That is, I describe a simple technique for separately and simultaneously seeing (and measuring) the monochromatic perception of the rods and the chromatic perception of the cones. I detail a number of experiments we conducted to validate this claim and then made use of this method of separated perception to directly measure the relative chromatic latency of cone-mediated vision. It is this observed and ordered chromatic latency that can be directly employed to dynamically read out the color information from the cones as generated by the proposed model.

This cone spectrometer model is inherently a dynamic one in the sense that it depends directly on temporal changes in cone input, as typically produced by the involuntary saccadic movements of the eye, to encode the distributed color information generated by the cone structure. As a dynamic model, it directly explains the phenomena of subjective color, the invocation of apparent color using only time-varying black and white illumination. A classic device used to demonstrate these subjective colors is Benham's Top and there has hitherto been no way to explain its phenomenology in terms of the classical, static three-cone model of color vision.

I also address the issue of color defective vision and suggest that the model provides a simple way to understand how the most common forms of such "color blindness" might be simply explained by physiological "mistuning" of cone spectrometer operation. Understanding how the cones might be mistuned in this way may provide an avenue for an effective therapy to ameliorate a defect of vision that affects nearly ten percent of the male population. The model makes some specific predictions about how this mistuning might be observed through some simple measurements that could verify or invalidate the proposed mistuning model for color blindness.

I would like to thank some of my collaborators who participated meaningfully in bringing the current work to light. In particular, anatomist Dr. Bessie Borwein and physicist Dr. James McGowan, both of whom I worked with while at the University of Western Ontario in the 1970's, were helpful with anatomical studies of the retina and its cone photoreceptors and in discussions about the cone spectrometer mechanism. The work presented here on the separation of rod and cone perception and the use of that work to measure the chromatic latency of color vision was conducted with Dr. George Caudle and Dr. Nancy Medeiros in the 1980's while at the Pennsylvania State University, York. I would also like to thank Dr. Nancy Medeiros for many useful suggestions and corrections in the review of this manuscript.

Any errors or omissions in this volume are entirely my own responsibility. I hope some readers will find this volume to be enlightening, it will certainly be controversial. Most importantly, from a personal point of view, is that after many years, this work will have finally gotten out in a published and accessible form. I can only hope that the general reader and the experts in the field suspend their belief of the "obvious facts" about the three-cone model of color vision long enough that they will fairly weigh the good, the bad and the ugly of what is addressed here. The claim is not that this is the complete and finished story, but rather only a beginning that may lead to a more complete understanding of human color vision.

John A. Medeiros
2006
john.medeiros@mac.com

# CONE SHAPE AND COLOR VISION:
## UNIFICATION OF STRUCTURE AND PERCEPTION

---

# Chapter 1          Introduction

The subject of human color vision, and particularly the question of just how the small and precisely formed cone receptors of the human retina mediate the resolution of the visible spectrum into the different perceived colors, has long been a matter of wide general interest and intense scientific research and debate. Despite this, many important questions remain unanswered and there is not yet a comprehensive model of the process that fully explains the myriad aspects of color vision phenomena.

In part, of course, this is a reflection of the complexities of a human subjective perception, involving on the one hand, still not understood information processing functions in both the peripheral and higher processing centers of the nervous system and, on the other hand, optical processes in small, complex and fragile receptor cells. Many different suggestions have been advanced over the years to explain how the receptors permit the resolution of color. A few of these have, at one time or another, competed with more or less success for attention as the correct model of the process.

The dominant theme about which the study of color vision and the associated model building have centered has been the trichromacy of metameric matches. The fact that any given illuminant in one half of a field can be matched by a combination of three primaries in the other half of the field defines this concept of trichromacy. That the infinite dimensional space necessary to describe the possible physical composition of an illuminant is, at least to a first approximation, reduced in the perceptual system to a representation of just three dimensions is a key aspect of human color vision and is the basis of the science of colorimetry. This characteristic clearly implies that at

1

some level in the visual system the spectral information incident at the retina is delimited by its encoding through three "channels" (Brindley, 1960). The exact nature of these visual "channels" is, however, rather enigmatic.

Trichromacy itself is but one aspect of a complex phenomenon; a model of the basic process by which the receptors discriminate colors must encompass a good deal of other experimental data including that on the structural features of the receptors and their organization in the retina, as well as the many steady-state and dynamic characteristics of the perception. One may pose the questions: what is the evidence dictating the nature of the basic process, and is the currently accepted model of color vision consistent with this evidence? That model is, of course, that the trichromacy of color vision is a result of the existence of three distinct cone populations, each with their own spectral sensitivities – the three-cone model of color vision.

While we will examine some of the evidence bearing on the key aspects of how human color vision works and its implications, it is not the purpose of this volume to exhaustively review the difficulties with the traditional formulations of trichromatic color vision theory. This has previously been carried out by a number of authors, including Balaraman (1961) who reviewed the historical development of the trichromatic theory, and Sheppard (1968), who critically examined the experimental foundations of human color perception. These reviews have concluded that the simple three-cone paradigm alone is not adequate to explain the phenomena and that fundamentally new directions are necessary. While these reviews are rather dated, the developments in the field over the succeeding decades have not fundamentally altered the validity of their conclusions.

# Why this Book?

The purpose of this volume is best described as an exploration of an alternative basis for understanding human color vision: using a very simple principle whereby the physical structure of the cone color receptors themselves can spatially separate light according to its wavelength, each cone acting as a miniature spectrometer. A role in color vision for this physical process, applicable to small tapered optical fibers, is not contradicted by current understanding of receptor function and physiology, and this concept will be seen to lead to a straightforward explanation of many different aspects of human color perception and to offer a real hope of yet reaching the goal of a comprehensive accounting of a seemingly complex phenomenon.

Original research is presented here, in which this color separation effect is directly demonstrated in optical fibers. The spectral dispersion observed in these fibers is precisely the mechanism we believe is present in the retinal cones. In addition, new experimental results are presented in which we directly observe the separate perception by the rod and cone photoreceptors of the eye. We use this separated perception by the photoreceptors in a novel method to directly measure the relative latency of color perception as a function of wavelength. These latency measurements are subsequently used to show how naturally occurring saccadic eye movements can convert the cone spectrometer effect into a color code for perception. Taken together, the model presented here along with these experimental results can form the basis of a new and comprehensive understanding of human color vision – one that is not contradicted by the available evidence and provides a more logical and connected way of understanding color perception.

Before undertaking this program of critical examination of the mechanism of human color vision, we first provide a bit of context on the eye and its workings.

# Eye Structure, the Visual Process, and Color Vision

In order to establish a basis for our discussions of human color vision, we first provide some general background on the overall human visual process including some information about the eye, retina, and photoreceptors. Of course, such broad topics fill entire volumes in their own right and it is only our purpose here to briefly sketch some background and a few salient details. Most of the information in the remainder of this introductory chapter will be quite familiar to specialists, and is provided only for the more general reader.

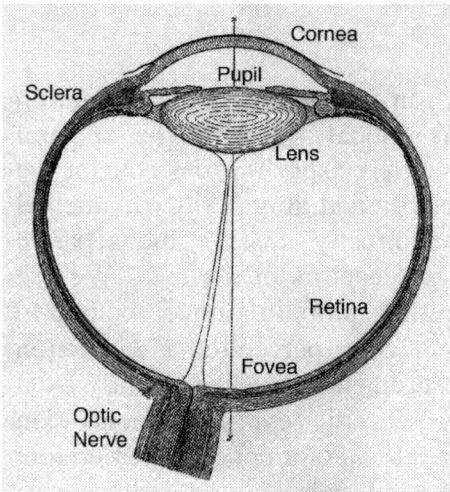

**Figure 1. Cross sectional schematic view of the human eye.**

The overall structure of the eye is schematically illustrated in cross section in Figure 1. Light enters and is refracted first through the transparent cornea, then the variably sized pupil and then the deformable lens before an inverted image of an external scene is focused on the photosensitive retina at the back of the eye. The tough, white outer coating of the eye is the sclera and its interior lining is the choroid, containing an extensive network of blood vessels. The photosensitive retina lining the inside back of the eye has a specialized area, the fovea that provides the most detailed vision at what corresponds to the center of fixation and best vision. The very center of this foveal area is known to contain only cone photoreceptors and the number density of rod photoreceptors increases at increasing retinal eccentricity. The foveal area corresponds to approximately one-degree total width in the visual field. The area of the retina serving central vision and extending out a few degrees from the fovea is known as the macula. The macula is overlaid with a yellowish pigment (the macular pigment) and is often distinctly visible as a differently colored area in photographs of the retina taken through the pupil (fundus view). Such views also reveal the optic nerve head through which the signals generated by photoreceptors are relayed to higher processing centers in the brain.

A fundus photograph showing a view to the back of a human eye is shown in the left panel of Figure 2. In this view, the optic nerve head is visible on the left and the foveal area in the center of the darker pigmented macula to the right of center. The blood vessels feeding the retina are seen emanating from the optic nerve head location and disappear as ever smaller capillaries nourishing the central area of the retina. The very central foveal area itself is seen to be avascular with no blood vessels present.

The right panel of Figure 2 is the same kind of fundus photograph, of the retina of another primate, a monkey. Note the remarkable similarity of the two views. This similarity is more than skin deep. Numerous anatomical, physiological, and psychophysical studies

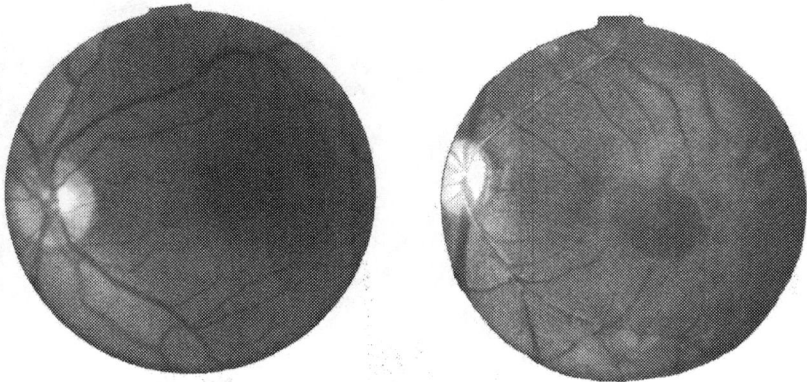

**Figure 2. Fundus photographs of the human retina (left) and monkey retina (right).**

have shown extensive similarity between the physical structure and function of the eye in man and monkey. This similarity is used to advantage in conducting studies on monkeys that could not be done directly on a human subject. The monkey (primate species, in general) is widely used as a surrogate model for the human eye.

The retina itself is a direct outgrowth of the brain and is one of the most complex and specialized pieces of tissue in the human body. There are a number of distinct cell layers comprising the retina. This complex of layers is sketched in Figure 3. In the orientation of this diagram, light is incident on the retina from above. Evidently, light passes through many of the nerve fiber layers before reaching the photoreceptor layer at the bottom of this view. In the living eye, the refractive index of all of these retinal layers are very similar so that there is little optical discontinuity for light passing through the retina before reaching the photosensitive layer. That is, these layers are

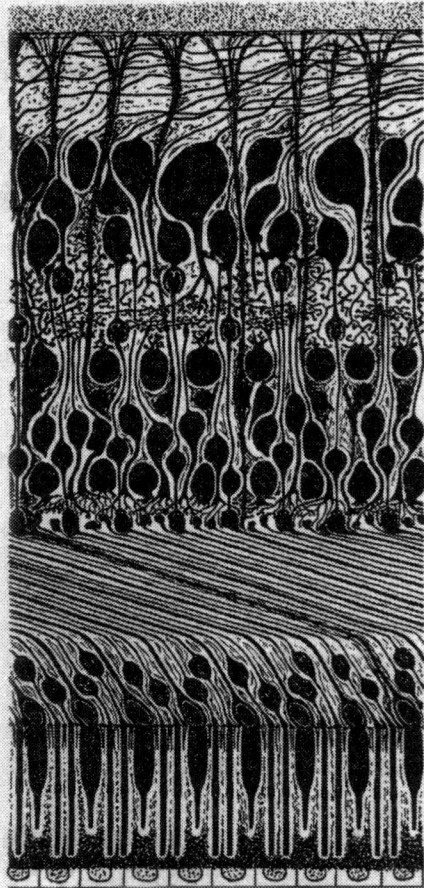

**Figure 3. Schematic illustration of the complex of layers making up the retina. In this view, light enters the retina from above, passing through the various cell layers before reaching the photoreceptors.**

highly transparent and do not interfere with light reaching the receptors.

The photoreceptors themselves consist of two distinct receptor types – the rods and cones. The rods are optimized for low light level operation (scotopic conditions) and mediate night vision. Rods provide only luminance information (black and white vision) and do not participate in color vision. The cones are optimized for operation at high light levels (photopic conditions) and provide color vision. Both receptor types have a similar anatomical plan and consist of two distinct segments. Light first passes through the inner segment, which contains numerous mitochondria that provide the energy required to sustain the phototransduction process. At its entrance end, the inner segment attaches to the bipolar cells that begin the processing of the detection information from the receptors. At their exit end, the inner segments couple the incident light into the photosensitive outer segment portion of the receptor cell. The outer segment contains a stacked array of disks (in the case of the rods) or a folded lamella sheet (in the case of the cones) that contain a packed array of photopigment molecules. The rod outer segments are cylindrical throughout their length and the cones are tapered, being broader at their entrance end where they are coupled to the inner segment and narrowest at their distal, terminal end.

It is well known that the distribution of rods and cones in the retina is highly non-uniform. The central foveal area contains only cones, which have their peak distribution there and decrease rapidly in number density out from the fovea and then steadily out to the *ora serata*, the extreme outer margin of the retina. The rod distribution starts from zero at the foveal center and increases with increasing eccentricity to a peak at around $20^\circ$ peripherally after which it decreases steadily out to the margin of the retina. These receptor distributions are illustrated in the classic receptor counting studies by Østerberg (1935) shown here as Figure 4 (see also Curcio, Sloan, Kalina, and Hendrickson, 1990).

This distribution accounts for the well-known observation that the most sensitive night vision is obtained through averted vision where one places the scene to be viewed in more peripheral portions of the retina. This is commonly observed at night in viewing a starry sky. Unless the star is very bright, a star visible in a peripheral view will be seen to disappear if one looks directly at it so as to place its image directly on the less sensitive rod-free foveal area.

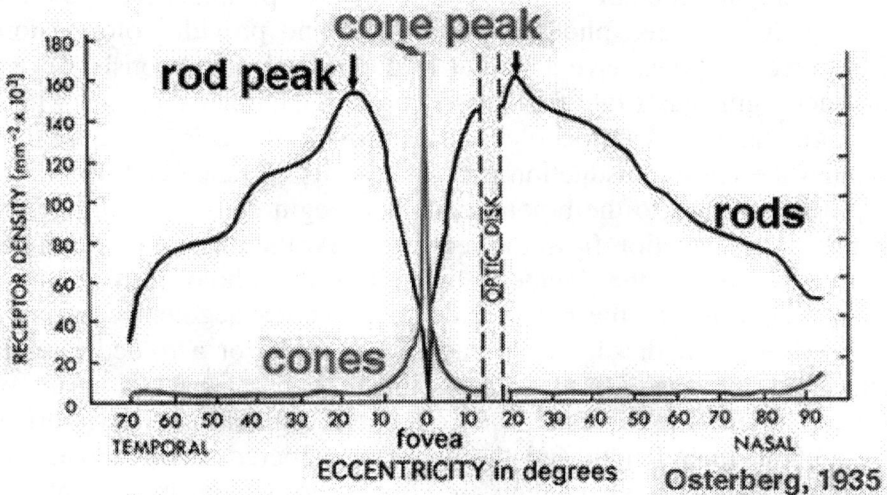

**Figure 4. Receptor density distribution in the retina.**

The receptors send their detection signals along a complex of nerve cells that both provide local processing and relay the information to the higher cortical areas of the brain. They first connect with the bipolar cells, which then connect to the ganglion cells. These ganglion cells eventually aggregate together to form the optic nerve and pass out of the eye at the optic nerve head and onto the brain. Within the retinal nerve layer there is also a complex of horizontal cells (at the photoreceptor – bipolar cell junction) and amacrine cells (at the bipolar – ganglion cell junction) that interconnect in separate layers within the plane of the retina and are involved in locally modifying the generated nerve impulses before the visual signals are passed on to the higher processing centers.

8

The receptors themselves are arraigned in a close-packed array at the back of the retina. This retinal mosaic can be seen in the series of photomicrographs shown as Figure 5, consisting of a series of electron micrographs of cross sections through the photoreceptor layer (near the central foveal area) in the monkey retina. These photographs were taken during some of our anatomical studies at the University of Western Ontario (Borwein, Borwein, Medeiros, and McGowan 1980). The close-packed nature of the photoreceptor matrix is evident along with the gradual changes in that matrix as one proceeds out from the central area.

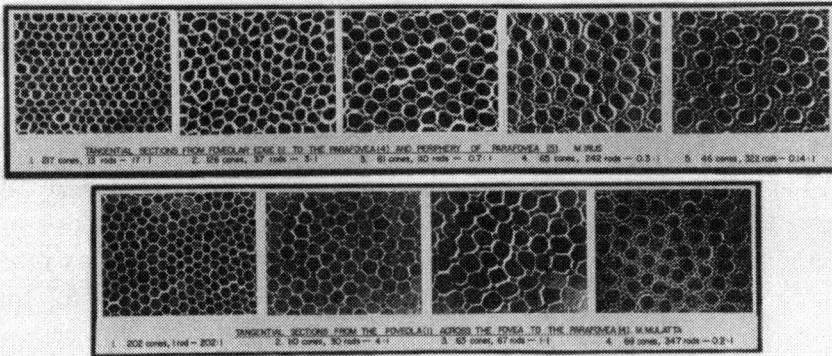

**Figure 5. Photoreceptor array in the central retina in two monkey species – Macaca irus top and Macaca mulatta below.**

The actual process of vision is initiated by a photodetection in the photoreceptors. This occurs only after the absorption of light by the photopigment resident within the so-called outer segment of the receptors. The general structure is essentially the same for all photopigments, whether resident in rods or cones. The photosensitive pigment (known as rhodopsin in rods) consists of a two-part structure. One portion is a large apoprotein opsin covalently linked to a smaller conjugated photosensitive chromophore, 11-*cis*-retinal, an aldehyde derivative of vitamin A. Current understanding is that the retinal portion is universally present, in the same form, in all visual photopigments. The opsin portion, however, is apparently different in rod and cone photopigments. Evidently, these differences modify the basic absorption spectrum of the photopigment.

The absorption of light triggers a physical transformation (isomerization, or shape change) in the retinal portion of the pigment. This in turn initiates a cascade of chemical processes that activates a G-protein and eventually results in a release of cGMP that closes an ion channel in the photoreceptor wall. This closure of the normally open channels causes a decrease in current from its dark, steady state value and constitutes the signature of the light absorption event. The structure of the photosensitive retinal molecule is shown in Figure 6 in its two states: prior to and just after the absorption of light. The light absorption initiates a rotation about the chemical bond arrowed in the figure that straightens the long side chain portion of the retinal molecule from its 11-*cis* to its all-*trans* state. The transformation to the all-*trans* state initiates the release of the neurotransmitter to signal light detection. The shape change of the retinal molecule is very fast and occurs on a nanosecond time scale. The nerve conduction processes this initiates occur on a millisecond time scale as is more characteristic for nerve conduction times. Subsequent to this sequence of events, physiological processes in the photoreceptors reset the retinal back to its 11-*cis* state and the pump is primed once again for participation in another detection event and the continuation of the visual process.

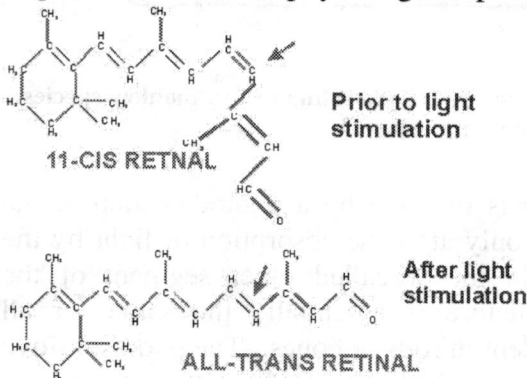

**Figure 6. Retinal isomerization on light absorption**

The sensitivity of the retinal receptors to light across the spectrum has been measured and reported many times in the literature. While there is considerable variation

10

from individual to individual in measurements of the exact values of this spectral sensitivity, a representative set of these measurements is shown from the work of Wald (1949) in Figure 7. These are sensitivity measures for the dark-adapted human observer. The cone and rod sensitivities are here plotted on the same scale. (Note that these are often plotted differently in the literature with the cone and rod maximum sensitivities both being set to unity which gives a highly misleading representation of the relative sensitivities of the receptors.) The plot is logarithmic in relative intensity so that the difference in sensitivity for the two-receptor types at the shorter wavelengths is much greater than it first appears on the plot. The cones and rods apparently have nearly the same sensitivity at the longer wavelengths (red) and the cones are two to three orders of magnitude less sensitive at the shorter wavelengths (towards the blue end of the spectrum).

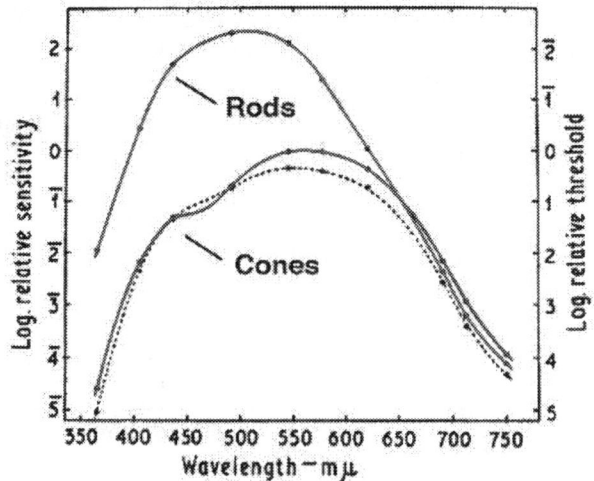

**Figure 7. Relative sensitivity of the two receptor types. The two cone sensitivity curves shown are for two different regions in the central retina (Wald, 1949).**

While it is still the case that no photopigment but rhodopsin, *presumed to be present only in rods*, has ever been extracted from the primate retina, it is equally evident that there must be other photopigments present in the retina as well, presumably cone photopigments. The genes encoding the opsin portions of several such pigments have been identified in the molecular genetic studies pioneered by Nathans, Thomas, and Hogness (1986) and others. These clearly indicate the existence of multiple pigments in the retina.

However, we will contend here that it is perhaps not so clear just what role these pigments play in the overall mechanism of human color vision.

A critical characteristic of human color vision is the color discrimination function, determined by measuring the discernable change in wavelength for an observer to tell that two colors are different. The standard technique for measuring this discrimination function is to present a two part field to a subject, one half of which is illuminated by light of wavelength $\lambda$ and the other half with an equal intensity of light of slightly different wavelength, $\lambda+\Delta\lambda$. One then plots the just discernable change in wavelength ($\Delta\lambda$) as a function of wavelength for the visible spectrum. Such a typical color discrimination plot is shown from the classic work of Wright and Pitt (1934) as Figure 8.

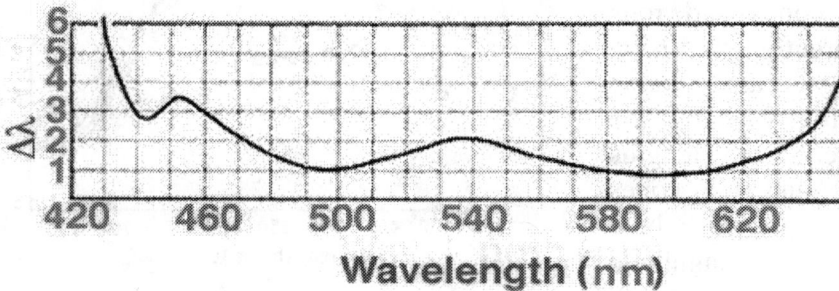

**Figure 8. Classic wavelength discrimination curve for human color vision (Wright and Pitt, 1934).**

A number of observations can be made about this plot. First, the discrimination fails at both ends of the visible spectrum as the change in wavelength required for the subject to detect a difference increases without bound at the extreme ends of the operating range of vision from about 680 nm at the red end to about 420 nm at the violet end. Second, there are two broad minima (best discrimination) at around 590 and 490 nm. These are presumably due the details of how color signals are processed in the visual system. The best discrimination of the visual system at around 590 nm, for example, corresponds to the location of unique yellow where just slightly longer wavelengths are perceived as reddish and just slightly shorter are perceived as greenish. A third feature of note is the very abrupt relative minimum in discrimination at around 440 nm. This corresponds to the perception of violet where as one starts from a blue wavelength and tunes to shorter wavelengths, it appears as if red is being added to the blue to form violet and resulting in the similarity of violet and an actual blue and red mixture, purple. This is commonly referred to as the "closure of the color circle", and there is no obvious, self-consistent explanation of this phenomenon in current color vision theory. We will return to this issue subsequently.

It is well known that the trichromacy of human color vision can be represented in different ways. One such way is by specification of the intensity of three different primaries, a red, a green, and a blue, for example. Another is to specify the value on the axis of three opponent color values (the Herring model, c.f. Hurvich and Jameson, 1957) such as Red-Green, Yellow-Blue, and Light-Dark (luminosity). A third way to specify this trichromacy is in terms of hue, saturation, and intensity. We discussed the hue dimension above in terms of the color discrimination function. For the saturation dimension, researchers typically measure this in terms of colorimetric purity, $P_c$, or difference from white. Measurements of this purity function were pioneered by Purdy (1931), Wright and Pitt (1937) and Priest and Brickwedde (1938). Experimentally, one can present to a subject a two-part field, one part white and the other white with a component of a spectral color added to it. The amount of the spectral component is varied, while keeping the total brightness of both fields the same, until the subject can tell that the part of the field with the chromatic component is different from the pure white part of the field. Quantitatively, defining $L_w$ and $L_c$ as the luminosity of the white and colored components, respectively, colorimetric purity is defined as:

$$P_c = L_c/(L_c+L_w).$$

Typical experimental measurements of colorimetric purity are shown in Figure 9. This is a plot of data from the work of Kraft and Werner (1999) who studied the effect of aging on the perceived saturation of colors. The plot is a measure of the colorimetric purity for older (lower curve) and younger (upper curve) subjects showing the function $\log(1/P_c)$ as a function of wavelength. This kind of V-shaped function is what is typically observed for people with normal color vision. The minimum in the curve, characteristically near the unique yellow hue around 580 nm, indicates the color that is least different from white. For the $\log(1/P_c)$ function used in the plot, a smaller number means it is less different from white and more of the colored illuminant must be added to tell that a chromatic component has been added to the white. A large value means the color is more different from white and very little of that monochromatic light needs be added to white to distinguish it from a pure white. Thus reds and especially blues have a high chromatic content and are easily distinguished from white while yellows are most like white. We will return to the subject of colorimetric purity and the saturation of colors in Chapter 5 in connection with a discussion of operational characteristics of the proposed cone spectrometer model for color vision.

**Pc Discrimination Data (Kraft & Werner, 1999)**

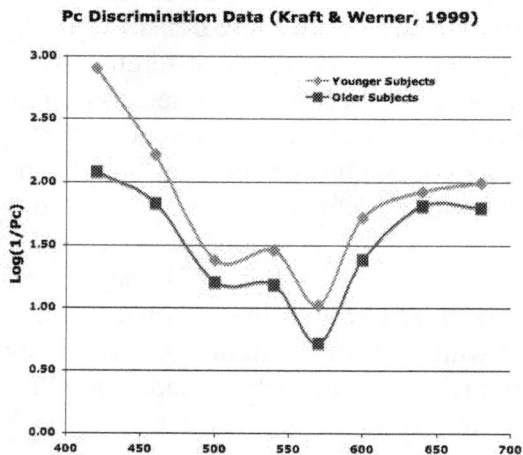

**Figure 9. Colorimetric Purity data for two groups of subjects.**

# Chapter 2     Models of Human Color Vision

## The Standard Model – What's Right and What's Wrong?

Thomas Young (1802) was among the first to recognize that the limited dimensionality of perceived colors was a result of the limitations of the sensory retina rather than of the domain of physical wavelengths. Since the time of his original suggestion, the most widely held model of human color vision has been some form of multiple receptor model. The assumption has commonly been that the three "channels" of the limited dimensionality of color vision are due to the existence of three types of cone receptor, each with its characteristic spectral sensitivity, most commonly assumed to be defined by the absorption spectrum of the one photopigment it contains. Thus the available spectral information is pictured as being reduced to a trichromatic representation in the very first stage of the visual process. Any one cone type then responds indifferently to the different physical wavelengths except that it responds with maximum amplitude, for a given intensity of light, to a particular portion of the spectrum. The color percept must then be synthesized in the subsequent neural network by inter-comparison of the response amplitudes of the various cone types.

While the efforts of many of the early researchers in color vision were directed toward identifying the spectral sensitivity of the cone types by looking for a unique set of three "primaries" (in color matching experiments) which would correspond to the responses of the cone types, it became clear (as was recognized by Helmholtz as early as 1852) that there is no such unique set of primaries for the representation of the possible spectral colors. There is an infinite set of possible "primaries", and an infinite set of possible trichromatic coordinates, all interchangeable through linear transformations.

15

## Multiple Cones or Multiple Pigments?

The questions of the exact number, distribution, and spectral sensitivity of the human cone photopigments remain enigmatic. Despite the relative facility with which the rod photopigment rhodopsin (assumed to be present *only* in rods) has been extracted from the retina, the identification of the three cone photopigments <u>by extraction</u> has yet to be accomplished (Sheppard, 1968; Criscetelli, 1972). The differential bleaching techniques used in the indirect methods of reflection densitometry (Baker and Rushton, 1965) and single cone microspectrophotometry (MSP) (Marks, Dobelle, and MacNichol, 1964; Brown and Wald, 1964) give results, which suggest multiple pigments although the data are not unequivocal. The MSP technique involves the focusing of a small beam of light (of adjustable wavelength) on a very small area of an excised piece of retinal tissue (the illuminated area is intended to encompass but a single cone receptor). The light reflected back from the photoreceptor area is then measured as a function of the incident wavelength. After this scan, the retinal tissue is then exposed to a bright light intended to bleach away any of the resident photopigment. The reflection scan is then repeated on the bleached retina and the difference in the two scans is interpreted as the absorption spectrum of the resident photopigment of the receptor.

The results of single cone MSP on the primate retina were widely hailed at the time of their announcement as providing the long awaited demonstration of the necessary cone pigments, although the tentative nature of the results was generally appreciated at the time. Despite the fact that minimal further results have been produced, the experiments are now often cited as being definitive and permitting the dismissal of any but the traditional Young-Helmholtz three-pigment formulation of color vision theory. The passage of time has not, however, altered the original results; caution is still called for in their interpretation. Consider the following points:

1. The number of individual receptors for which results were reported is too small for a meaningful analysis. Marks, et al., (1964)

16

reported observations on two human cones and eight monkey cones. Brown and Wald (1964) reported results on four human cones.

2.    No true red receptor was found.  The absorption maximum of the longest wavelength receptor peaks in the *yellow* region of the spectrum.  In addition, the data in the original reports suggested "the possibility that the 'red' receptor may contain red and green pigments coexisting in a single cone" (Marks, et al., 1964).

3.    The experiments have been criticized on fundamental grounds of procedure by Liebman (1972).  He notes that the rapid post-mortem degeneration in the neural layers of the primate retina and the attendant scattering of light severely limit the results. Furthermore, use of the high numerical aperture objectives used in the measurement optics only insures that most of the light in the test beam passes outside the individual receptor and that the consequent bleaching of the surrounding rods can influence the observations (Enoch, 1966).  Liebman concluded that the data alone cannot be regarded as accurate to better than 20-30 nm and notes: "Unfortunately almost none of the original data has ever been shown in reports on the primate pigments, and no mention has been made of the unacceptable experimental conditions that must have been tolerated."

4.    Riggs (1967) plotted the MSP and the reflection densitometric results on a single graph, which we reproduce here as Figure 10.  As is evident, the cone photopigments identified by the techniques do not fall into three clearly defined classes.  Instead, they form an essentially continuous distribution of broadly overlapping curves, with a slight break towards the blue end of the spectrum, c.f., Enoch (1972).

Of course, even if the single cone MSP results were less equivocal, proof of multiple cone pigments would not in itself constitute proof of the traditional Young-Helmholtz model of color vision.  Multiple cone pigments are a necessary condition for such a model, but their demonstration is not alone sufficient to exclude other possibilities (for example, that the multiple pigments played a secondary role in optimizing the efficiency of some entirely different mechanism).

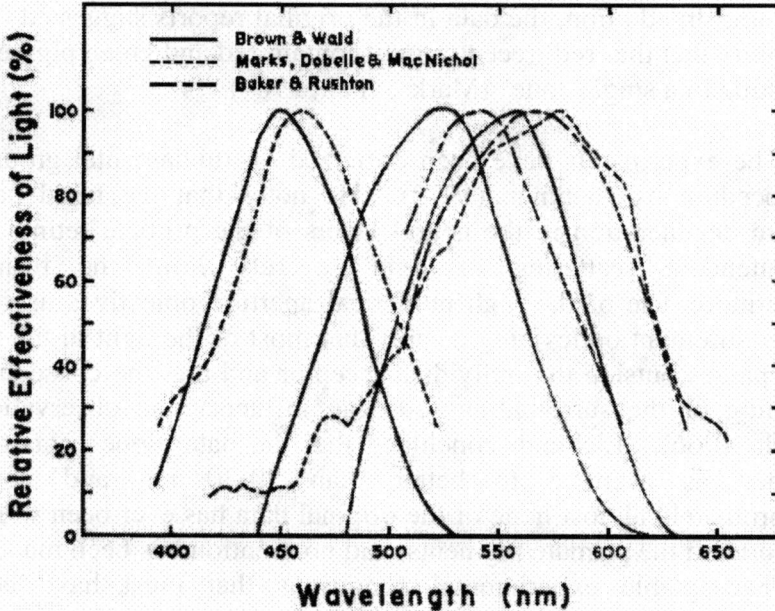

**Figure 10. Plot of cone photopigment curves from MSP and reflection densitometry measurements. (Riggs, 1967)**

## Molecular Genetics of Cone Pigments

Direct support for multiple photopigments in the retina has been provided through the molecular genetic investigations pioneered by Nathans and others (Nathans, Thomas, and Hogness, 1986; Neitz, Neitz, and Jacobs, 1993; Sharpe, Stockman, Jägle, and Nathans, 1999). These ingenious studies have circumvented the difficulty of directly detecting the cone photopigments in the cones themselves by identifying and characterizing the genes that code for them. The results of this work have been widely interpreted as finally closing the issue as to the source of the trichromacy of human color vision – that is, clear evidence of three cone photopigments means three types of cones and the problem is finally solved. However, the issue is substantially more complicated than this cursory judgment would warrant. Molecular genetic studies have revealed not just three unique cone photopigments, but rather a range of variants of several photopigments with a corresponding range of absorption spectra maxima. Moreover, the presence (or absence) of one or more of these pigments does not correlate directly with the observed function (or deficit) of color vision. There is evidence that subjects with normal color vision may have more than three cone photopigments (Neitz, Neitz, and Jacobs, 1993; Neitz and Neitz, 1995) and there is evidence that some subjects with anomalous color vision are entirely missing one of the photopigments, but yet have trichromatic color discrimination (Neitz, Neitz, and Jacobs, 1989; Neitz, Neitz, and Grishok, 1995).

The evidence from the molecular genetic studies play such a critical role in current understanding on human color vision that it clearly warrants more detailed attention in the present context. As discussed in Chapter 1, the photopigment molecules of the retina (whether resident in rods or cones) consist of two interlocking components, a basic chromophore (11-cis retinal) that undergoes a physical shape transformation under light absorption and a transmembrane opsin (or apoprotein) to which the chromophore is covalently linked and whose spectral sensitivity it modifies. All opsins consist of seven helices (a heptahelical protein) linked by intra- and extracellular loops. Evidence suggests that these opsin helices enclose the retinal molecule in a central binding pocket. The binding site for the retinal in all opsins is located in helix 7. When the photopigment molecular components are separate from each other, the opsins have their optical absorption maximum in the ultraviolet at wavelengths below 300 nm and the retinal absorbs maximally at about 380 nm. It is only when bound together that the complex exhibits a characteristic broad absorption maximum in the visible spectrum. The exact value of that absorption maximum depends on the structural details of the opsin molecule.

19

The approach in molecular genetic studies is to identify and characterize the genes responsible for the different opsin complexes that are identified as resulting in different photopigments with characteristic absorption spectra. Paraphrasing the excellent review on photopigment genetics by Sharpe, et al. (1999): four separate genes encode the opsins of the S-, M-, and L-photopigments and of the rhodopsin rod photopigment. The genes encoding rhodopsin and the S-photopigment exist as single copies: the rhodopsin gene on the long or q-arm of chromosome 3, within a cytogenetic location between 3q21.3 and 3q24 and the S-photopigment on the q-arm of chromosome 7 between 7q31.3 and 7q32 (Nathans, et al., 1986). The situation for the genes encoding the L- and M-photopigment is more complicated. They reside on the q-arm of the X-chromosome at location Xq28 within a head-to-tail tandem array, which may contain as many as six gene copies. The L-pigment gene is, in general, present only as a single copy and precedes the multiple M-pigment genes in the array.

The molecular genetic studies have revealed evidence for cone photopigments with absorption maxima at 560 nm (L), 530 nm (M), and 425 nm (S) (Nathans, 1999). In addition a *multiplicity* of slightly altered variants of the L and M genes which evidently give rise to photopigments with slightly different absorption maxima have been found for subjects with normal as well as with color defective vision (Neitz and Neitz, 2000). Indeed, this 'embarrassment of riches' in the number and type of pigment genotypes leads to the view, characterized by Neitz and Neitz (2000) in their review, "Molecular Genetics of Color Vision and Color Vision Defects":

> "The molecular genetics of color vision has turned out to be much more complex than originally suspected. This complexity derives in part from the fact that the red and green opsin genes are adjacent to one another and they are about 98% identical. It seems that during human evolution, because of their close proximity and high similarity, the red and green genes were subject to frequent homologous

recombination. This, perhaps in conjunction with relaxed natural selection against color vision defects in civilized humans, has given rise to a great deal of variability in the red and green photopigment genes. The rearrangements have included duplications of the red and green genes so that most people have extra pigment genes. Individual X-chromosomes contain variable numbers of red and green genes arranged in a tandemly repeated array. Nevertheless, in the face of the unanticipated complexity, much progress has been made toward understanding the relationship between color vision genotype and phenotype."

This complexity is very apparent when the standard 'missing cone type' approach to explaining defective color vision is considered. In discussing the situation in regard to the molecular genetic observations on subjects with one type of red-green deficit color vision, Neitz and Neitz (2000) further observe:

"In contrast, there is an aspect of deuteranomaly in which the relationship between genotype and phenotype is not clear at all, i.e., most persons with deuteranomaly have M pigment genes and thus it is not understood why they do not have M cone function. This is probably the most important unanswered question concerning the molecular genetics of color vision defects. There is evidence that persons with deuteranomaly lack an M cone contribution to vision because they lack both functional M cones and expressed M photopigment. However, what causes this loss is uncertain."

There are additional complications. In a report on "Numbers and ratios of visual pigment genes for normal red-green color vision" (Neitz and Neitz, 1995) observed:

"Red-green color vision is based on middle-wavelength- and long-wavelength-sensitive visual pigments encoded by an array of genes on the X chromosome. The numbers and

21

ratios of genes in this cluster were reexamined in men with normal color vision by means of newly refined methods. These methods revealed that many men had more pigment genes on the X chromosome than had previously been suggested and that many had more than one long-wave pigment gene. These discoveries challenge accepted ideas that are the foundation for theories of normal and anomalous color vision."

In yet another report, "Analysis of fusion gene and encoded photopigment of colour-blind humans", Neitz, Neitz and Jacobs (1989) sequenced a fusion gene (combined red-green pigment) from a red-green color blind subject and measured the spectral properties of the pigment it encoded. They observed that the "absorption spectrum of the photopigment was very similar to that of normal middle-wavelength-sensitive photopigment, even though about half of its DNA coding sequence seems to be derived from a gene encoding normal long-wavelength-sensitive pigment." They thus concluded: "These results indicate the regions of the X-encoded photopigment apoproteins that are responsible for differences in their spectral tuning, and imply that the striking variations in colour vision among anomalous trichromats of a particular type are not attributable to anomalous pigments with differing spectral peaks."

Even in terms of observers with normal color vision, this complexity in interpreting the results of the molecular genetics remains. In another study, entitled "More than three different cone pigments among people with normal color vision." Neitz, Neitz, and Jacobs (1993) compared the difference in the amount of red and green needed to exactly match a spectral yellow for color normal observers with their differences in photopigment genes. They found a serine/alanine polymorphism at amino acid position 180 of X-encoded pigments that could account for this type of color vision variation (in variation of the red-green mixture needed for a yellow match). However, they noted, "This amino acid change shifts the spectrum of the pigment produced by about 6 nm, a value that would predict a larger minimum color vision difference between individuals

22

than is actually observed." Their bottom line conclusion of this study was: *"This discrepancy can be explained if, counter to the Young-Helmholtz theory as the explanation of trichromacy, many people with normal color vision have more than three spectrally different cone pigments."*

In the standard view that each L- and M-photopigment is uniquely sequestered in a characteristic cone type, experimental observations by various techniques have apparently revealed a very large variation in the L:M cone ratio in normal retinas. In an experiment to test one hypothesis behind the origins of this variation, McMahon, Neitz, and Neitz (2004) describe how:

> "Each gene has a promoter, and upstream of each array there is a genetic element, termed the locus control region (LCR), that is required for the expression of both L and M pigment genes. During development, for each cell that has been determined to be an L or M cone, it has been proposed that the LCR acts as a stochastic selector, which chooses one gene from the array to be expressed. In this model, the L and M promoters compete for contact with the LCR in each photoreceptor. Theoretically, the promoter that, by chance, is the first to successfully form a stable and permanent complex with the LCR commits the cell to a lifetime of exclusive expression of its associated gene. Under this model, it has been suggested that nucleotide differences in the promoters influence their ability to compete in forming a complex with the LCR. Thus, normal variation in L:M cone ratio is predicted to be associated with nucleotide polymorphisms in the promoters."

In a very detailed experiment, they ruled out this possibility, noting: "the vast majority of cone ratio differences were not associated with any difference in the promoter sequence." They concluded:

> "To explain the high degree of cone ratio variation among normal males, the mechanism that determines whether a

cone is L or M must involve genetic elements that have a high degree of genetic polymorphism in the normal population. The results presented here indicate that there are additional genetic components of the mechanism which remain to be identified and incorporated into the present hypotheses."

Finally, we conclude this examination of the complexity of the standard interpretation of the molecular genetic results with the, apparently, remarkable results in a study by Neitz, Neitz, He, and Shevell (1999) entitled, "Trichromatic color vision with only two spectrally distinct photopigments". They examined and compared one type of red-green color defective subjects (protanomolous) with color normal subjects. They observed that "Whereas normal subjects have pigments whose wavelengths of peak sensitivity differ by about 30 nm, the peak wavelengths for protanomolous observers are thought to differ by only a few nanometers. We found, however, that although this difference occurred in some protanomolous subjects, others had pigments whose peak wavelengths were identical." They concluded:

"Genetic and psychophysical results from the latter class indicated that limited red-green discrimination can be achieved with pigments that have the same peak wavelength sensitivity and that differ only in optical density. A single amino acid substitution was correlated with trichromacy in these subjects, suggesting that differences in pigment sequence may regulate the optical density of the cone."

It is not our purpose here to suggest that results of the molecular genetic studies are not correct, but rather to emphasize the necessity to exercise caution in their interpretation. The resulting story on cone photopigments is indeed complex and it is not at all obvious that it supports the classical interpretation of the three-cone model of human color vision. The evidence is clear that there are multiple pigments in the cones, but it is not at all clear that they are sequestered with one unique L-, M- or S-pigment type in a respective L-, M-, or S-cone as required in the classical view.

In an attempt to find such a unique correlation, researchers have tried turning to a direct examination of the retinal mosaic through reflection densitometry. Reflection densitometry techniques have been vastly improved in recent years through optical improvements, including the use of adaptive optics techniques to illuminate retinal regions with the dimensions of individual cones (Williams, 1988; Elsner, Burns and Webb, 1993; Roorda and Williams, 1999). For example, Roorda and Williams (1999) have produced images by the technique apparently showing red, green, and (rarely) blue dominated reflectances from the cone positions. It is easy to conclude that such results provide the needed proof of multiple cone types. The situation is, again, certainly more complex than such a simple interpretation would warrant. In the reflection densitometry technique (with or without adaptive optics) the approach is to measure the light reflected back from the individual cone receptors before and then after the application of intense illumination intended to bleach any photopigment in the receptors. From the observed difference in pre- and post-bleaching spectra, one infers the absorption spectrum of any photopigment resident in the receptors. In the retinal mosaics published with this technique, pseudo colors of red, green, or blue are then applied to the individual cones in correspondence with the absorption spectrum observed.

Note that, this technique ignores the details of any waveguide effects that can modulate coupling to the cone photoreceptors or how these waveguide properties may affect backwards reflected light from within the cone. The technique probes the contents of the cone to some depth, but that depth depends on the precise details of the entrance conditions into the receptors. Furthermore, any light reflected would also depend on the details of how any resident photopigment is distributed along the length of the cone. That is, uniquely assigning a photopigment absorption spectrum to a specific cone assumes that only one photopigment is resident there and that, for example, we do not have the situation that one photopigment is present in the wide distal end of the cone and perhaps some other pigment is resident in the more narrow, distal end of the cone.

In any event, all of these methods have shown enormous variability in the apparent ratio of long-wavelength absorbing (L) to middle-wavelength absorbing (M) cones (and very few short-wavelength absorbing, S cones) from individual to individual. Despite this variation, color vision characteristics of the subjects differ little between these individuals - including their spectral sensitivities, hue discrimination functions, or the location of their unique hues. This has required the presumption of vastly different densities of the photopigments to maintain uniformity of function across these variations.

While it is not my purpose to exhaustively review the methodology of the techniques, nor even the apparent evidence for distinct photopigments supported by these results, there is no reason to dispute the existence of multiple pigments. We would, however, again take pause with the conclusion that evidence for multiple pigments is evidence for multiple cone types. That is, it is a separate matter to prove that there may be different cone pigments and then that these separate cone pigments are uniquely segregated, each in its own distinct cone type. Evidence for multiple pigments is not evidence for multiple cone types nor that three such cone types provide the basis for human color vision. It may be that multiple pigments enhance some other mechanism through a process where more than one photopigment might exist within a given cone and that the basis of spectral discrimination is the result of some other process.

A critical issue underlying the basic mechanism of human color vision is the following: where, and as a result of what feature in the visual system, is the limited dimensionality of color vision imposed? While a multiple receptor model offers a readily visualized scheme whereby the approximate trichromacy of metamerism may result, the available evidence does not rule out the possibility that the receptors could resolve the full spectrum and that only subsequently, as a result of limitations in the neural network reading the outputs of the cones, is the available information reduced to a limited dimensionality.

There are at least two major lines of evidence arguing against the traditional view that the partitioning of the spectral information into three "channels" occurs in the very first stages of color vision, at the receptor level. First, the sensory retina and its color receptors are simply not physically organized in a manner consistent with the multiple receptor concept. Secondly, demonstrably more information than such a model allows is present in human color vision function. We examine these two aspects in the next two sections, one on the local aspects of structure and perception and one on the dynamic aspects of color vision.

## Local Effects: Structure and Perception

There are two key aspects of retinal photoreceptor structure that should be considered: the structure of the cone color receptors themselves and the organization of their connecting output (bipolar) cells. The photosensitive (outer segment) portion of the color receptors of the eye is universally cone shaped. The conical shape, however, plays no role in the three-cone model of vision. Moreover, it is well known that all the cones within a given local region of the retina are essentially identical in size, shape, and morphology (Polyak, 1941; Biernson, 1966; Cohen, 1972; but see below in the section on blue cones). However, the physical form of the cones – their length and taper – does change gradually and systematically across the extent of the retina. That the cone receptors are physically locally uniform is not in itself usually held against the multiple receptor models, since the differences among cones are simply relegated to the molecular level in the form of different photopigments. It is, however, rather difficult to see, as Biernson (1966) has emphasized, how these differences have come about with no evident differences in the embryological development of the cones.

Not only do the cones within local regions of the retina appear indistinguishable in structure, but neither are any local variations in function observed. The so-called "blue-blindness" of the central fovea (Wald, 1967) does not constitute an exception to this statement since that variation is more closely associated with the rigid fixation necessary to probe the color function of the fovea than with the absence of blue receptors (Wright, 1971). Subpopulations of cone types with different spectral sensitivities have long been sought but have not been observed, either through comparisons of visual acuity in differently colored illuminants (Polyak, 1941; O'Brien, 1951; Enoch, 1967) or through probing the receptor matrix by moving a small spot of light over the retina (Brindley, 1953; Ditchburn, 1957).

The recent experimental results of Hofer, Singer, and Williams (2005) illustrate this point dramatically. They noted that it has been difficult to probe regions of the retina as small as individual photoreceptors because of the blur due to the eye's optics. They overcome this limitation by using an adaptive optics technique where they are able to present monochromatic flashes of light on the retina with a blur circle smaller than an individual cone. They were thus able to probe the retinal mosaic in a manner where they were able to elicit the sensations from a single receptor at a time. Under these conditions, the clear expectation is that they would find that their monochromatic stimuli (600 nm, 550 nm, and 500 nm) would elicit the corresponding sensation of red, green, or blue depending on which cone they stimulated. Instead, they essentially found that stimulation of any one cone by any one color could elicit virtually any color sensation, ***including the sensation of white***.

While they assumed these results must imply some complex interactions of the antecedent retinal circuitry, they none-the-less concluded that: "The diversity in the color response could not be completely explained by combined L and M cone excitation, implying that photoreceptors within the same class can elicit more than one color sensation." They further observed that: "This result implies that white sensations can result from excitation of cones of only one class. Apparently, then stimulation of cones containing the same photopigment can give rise to different sensations."

Hofer, et al. (2005) further observed that "The spatial grain of the cone mosaic is remarkably invisible in perceptual experience (Williams, 1990)." Given this general observation, they made the effort to drill down to the individual receptor level and noted: "But adaptive optics allows us to present stimuli on a smaller spatial scale than arises in normal perceptual experience, stimuli for which cortical circuitry had no opportunity to develop. Our experiments firmly reject the idea that excitation of all cones within the same class results in the same hue sensation." They further noted "Consequently, every cone of the same class cannot make the same contribution to cortical circuitry for extracting hue and brightness."

Another "remarkable but forgotten property of human color vision" (so-termed and investigated in detail by Mollon and Estévez, 1988) was first described by Tyndall (1933) in which wavelength discrimination at around 455 nm is actually observed to *improve* when a substantial amount of a white light desaturant is added. Quantitatively, it was found that wavelength discrimination improved by nearly 50% when as little as 20% of the illuminant was a monochromatic blue (while 80% of the illuminant was a de-saturating white). As Mollon and Estévez (1988) point out, the explanation of this result is less than straightforward (in terms of three-cone models) since "the quantum catches of both the shortwave and the middlewave receptors are altered considerably" with the addition of the desaturating illuminant. They concluded that: "Our results show that wavelength discrimination cannot be limited only by the rate of quantum catches in different classes of cones."

While this remarkable property of human color vision is not readily explained in any obvious way since it clearly depends on some sophisticated post-receptoral processing occurring in the eye, it is clear that this is an impossible result for three-cone models. Diluting the monochromatic stimulant in this way <u>must</u> reduce the ability of a three-cone mechanism to discriminate wavelengths. The uniqueness of the quantum catches by the individual receptor types would be reduced and wavelength discrimination must get worse – in contradiction to the experimental results.

Data on the dimensionality of human color vision for different regions of the retina and the corresponding neural organization in those retinal areas argue strongly against the traditional three-cone model. We compare here reported observations on the dimensionality of color perception across the retina with cell counting studies on the relative distribution of cones and cone bipolar cells (their outputs) throughout the human retina. These cell-counting studies (Polyak, 1941; Vilter, 1949; Missotten, 1974) reveal that there are three bipolars per cone in the central retina where color vision is most acute; that is, the afferent neural network is divergent (contrary to what is usually assumed in the literature). The ratio of cone bipolars to cones decreases for the intermediate regions of the retinal periphery to 2:1 finally to 1:1 in the far periphery.

It is well known that color vision is most highly developed in the central retina and falls off in resolution toward the retinal periphery (Weale, 1953; Moreland and Cruz, 1959; Boynton, Schafer, and Neun, 1964; Moreland, 1972; Wooten and Wald, 1973; Abramov, Gordon, and Chan, 1991 & 1992). While the details of color perception in the peripheral retina depend importantly on the brightness of the stimulus and on the size and adaptational state of the retinal area stimulated, studies reveal the existence of distinct zones of reduced color vision function (its dimensionality) across the retina. These zones of color vision are: trichromatic color resolution in the central region out to about 20 to 30 degrees, dichromatic in the intermediate region out to about 40 to 50 degrees, and essentially monochromatic out to the far periphery.

The correspondence between the bipolar/cone ratio and the dimensionality of color vision in the various retinal zones is very close, as is apparent in Figure 11 (see Sheppard, 1968, for further discussion of these results). This association strongly suggests a more central role for the bipolar cells in the processing of color information than is usually ascribed to them. Indeed, the clear implication of these results is that the three "channels" of color vision are located in the secondary layer of the retina, associated with the number of outputs of each cone. The association of the dimensionality of color perception with the number of physical outputs to which the cones are attached in turn suggests that the partitioning of the spectrum is occurring only after the basic detection events in the cones.

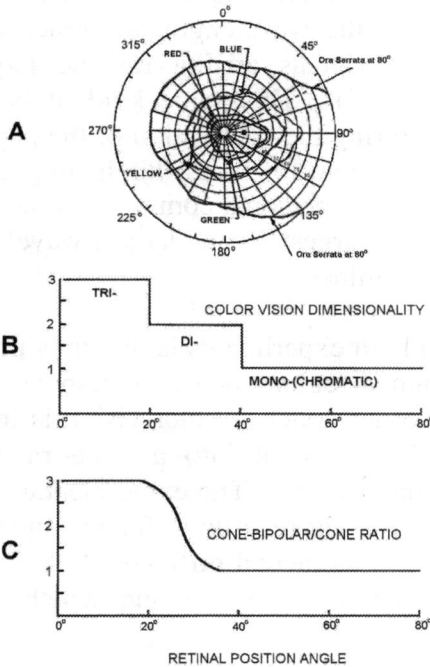

**Figure 11. Relationship of color vision quality and the ratio of cone bipolars to cones. (A) Typical pattern exhibited in perimeter measurement of human color vision. Red and green perception is lost essentially together at around 20° – 30°, blue and yellow perception around 30°-50° (Optical Society, 1960). (B) Schematic representation of perimeter measurements; the dimensionality of color perception is plotted as a function of retinal position angle. (C) Ratio of cone bipolars to cones across the retina as seen in the cell counting studies and displayed on the same scale as the dimensionality representation, B above.**

31

*John A. Medeiros*

# Dynamic Characteristics of Color Vision

In terms of receptor function, there is considerable evidence that the actual wavelength composition of light, not just its trichromatic coordinates, is important (c.f. Sheppard, 1968). For example, Riggs and his collaborators found that the magnitude of the electrical response of the eye to alternating lights of different wavelengths is a monotonic function of the wavelength difference of the alternated pair with the largest signals arising for the largest wavelength differences (Riggs, 1967). This same kind of behavior has been observed when monitoring the contraction of the pupil in response to a sudden change in the color of light illuminating the retina (Young and Alpern, 1976; Young, personal communication). The magnitude of the pupil response is largest for the largest wavelength differences in suddenly altered illuminants.

What is common to both experiments is that they probe the dynamic characteristics of human color vision, its response to time-varying stimuli. The three-cone model of color vision is inherently a static model. It offers little insight into a wide range of seemingly unrelated temporal phenomena. The experimental data itself is both contradictory and confusing regarding, for example, questions such as whether or not there are actual differential chromatic latencies in the electrical response of the retina and whether red or blue is perceived faster.

In part, this situation reflects the difficulty of measuring effects that depend only on differences in wavelength independently of brightness differences. Indeed, heterochromatic brightness matching is notoriously difficult and there is a certain arbitrariness to the definition of "equal brightness" of red and blue lights, for example. Chromaticity information and luminosity information are processed in different ways and with different speeds in the human retina. Thus, Weingarten (1972) observed different latencies for red and green light only when a hue-substitution technique was used; Kinney and McKay (1974) observed that longer processing times are required for chromaticity information than for intensity information by a pattern contrast measure; and it is found that intensity information and color information are lost in different ways under conditions of retinal image stabilization (Ditchburn, 1957).

The importance of the dynamic properties of color vision is underscored by the existence of the subjective color phenomena induced by intermittent achromatic illumination as, for example, in the well-known Benham's Top phenomenon (Cohen and Gordon, 1949). We will examine this effect in more detail in Chapter 6 when we take up the issue of temporal coding of color information. For now, it is sufficient to note that it is known that a specific temporal coding of flickering black-and-white patterns is both necessary and sufficient to evoke particular color percepts (Sheppard, 1968; Festinger, Allyn and White, 1971; Polizzotto and Perura, 1975). This temporal code for subjective colors is the same for *all* normal observers; while there are individual variations in the clarity and saturation with which the colors are seen, all agree on the spectral ordering of the colors induced as the phase of the time code is altered. The universal nature of this code indicates that the temporal effects are directly associated with the basic color perception mechanism itself.

It is clear that even under normal viewing conditions, without intermittent illumination, the dynamic properties of the receptors are of central importance to vision. When the retinal image is made truly static through the elimination of the small involuntary eye movements, visual perception, and particularly color vision, is rapidly and completely lost (Ditchburn, 1973). Theoretical understanding of human color perception is necessarily incomplete without an understanding of its dynamic properties, evidently an integral part of visual function.

In the multiple receptor framework one would naturally seek to explain the temporal color phenomena in terms of the different receptor types having different time constants. Such "explanations" have not been notably successful nor, in any case, is it obvious why apparently identically structured cones should have different time constants in the first place nor why these time constants are spectrally ordered in the same way for all observers.

A critical illustration of the fundamental difficulty with multiple receptor explanations of color vision is the breakdown of metameric matches (established under static conditions) when presented under dynamic conditions. This was demonstrated in a simple experiment performed very early in the Twentieth Century by H.E. Ives (1918). The significance of his very important result has subsequently been all but ignored (but see Cohen and Gordon, 1949 and Crawford, 1972). This, despite the fact that Ives set up his experiment so that the eye was used as a comparator, or difference detector, constituting a so-called Class A observation. This was conducted nearly forty years before Brindley (1960) first clearly drew the distinction between experiments which use the eye as a null detector (Class A) and ones which determine the particular sensation perceived (Class B) and stressed the importance of Class A experiments as true critical tests of theoretical understanding.

Ives compared the appearance of lights, which were metameric matches when presented statically, to their appearance under dynamic presentation as the images were moved across the retina. When the image of a slit is moved across the retina, the cones at the leading edge of the image are turning on as illumination first reaches them, and at the trailing edge cones are turning off as the image passes by. Ives combined a red and a green illuminant such that the combination appeared yellow. When the mixed yellow was moved across the retina, the leading edge appeared red and the trailing edge green. Ives then repeated this with the spectral yellow with which the compound yellow was metameric: *"The next point taken up was the behavior of the pure yellow, adjusted to be a subjective match with the compound yellow, and arranged to exactly take its place between the red and green. It was at once apparent that <u>pure yellow does not separate into red and green</u>. This fact is strikingly shown by arranging the slit so as to be all compound yellow, except a small portion of pure yellow. When stationary the slit appears alike throughout its whole length in brightness, hue and definition.*

*But upon moving the image sideways, or oscillating it, the compound yellow immediately broadens out and becomes ill-defined, the pure yellow remaining narrow and sharp"* (original emphasis).

This definitive result clearly contradicts the view that color information is partitioned at the receptor level. In that view, color differences are discerned by the differential sensitivity of cone classes, and the compound and pure yellows are a metameric match because the different cone classes are excited in the same proportion for the two "yellows". Thus even if the "red" cones and the "green" cones had different time constants, thereby explaining the resolution of the mixed red and green, the nature of the model requires that the pure yellow must similarly resolve into the component sensations, in contradiction to the experimental result. Ives' subjects also resolved a combined red and blue and he noted that while the phenomenon of color mixture evidently calls for a trichromatic mechanism, the result of his experiments *"calls for the additional complication of an antecedent transmitting process where colors retain their physical (spectral) individuality."*

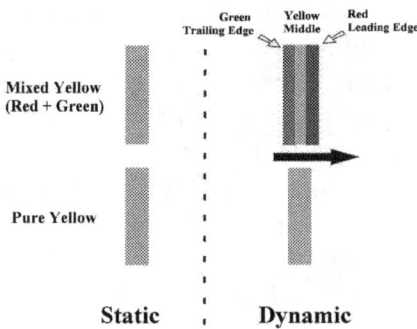

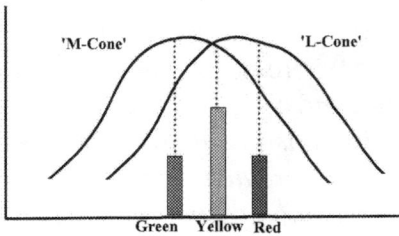

Figure 12. **Breakdown of static metameric matches under dynamic conditions and the quantal catch view in a multiple cone model.**

The observer's view in this experiment is schematically illustrated in Figure 12. This shows the appearance of a two-part yellow bar, one part a red and green mixture and the other a monochromatic yellow in both a static view

35

and a view under dynamic conditions. The "mixed" yellow and "pure" yellows are adjusted to be a metameric match under dynamic conditions. Under dynamic presentation of motion (in this case to the right) the "pure" yellow remains unaltered, but the "mixed" yellow broadens-out with a red leading edge, a green trailing edge, and a yellow middle. The three-cone mechanism of static metameric matching is also illustrated in the figure (where we show the presumptive contributions of only the L and M cones with their absorption spectrum overlaying the location of the respective red, green, and yellow illuminants). In the "pure" yellow case, absorption (quantal catch) by the L and M cones each have a particular value (perhaps equal). In the "mixed" yellow case, the quantal catches by the L and M cones are identical to the "pure" yellow case (as set by adjustment of the red and green mixture). Since the receptors – in the three-cone model – simply signal only the total quantal catch, their output must (by definition) be identical in the "pure" or "mixed" yellow cases. If the L and M cones have different time constants, giving rise to the leading red edge and trailing green edge for the "mixed" yellow when the bar of light is moved, the identical result must happen for the "pure" yellow. The fact is that it does not and this simple model is clearly wrong.

This resolution of mixed colors has apparently been observed and discussed, in a somewhat different context, in a recent study. Nijhawan (1997) describes an experiment where he also observes the decomposition of mixed yellow. In his words: "...*the visual system can decompose a 'yellow' stimulus into its constituent red and green components. A 'yellow' stimulus was created by optically superimposing a flashed red line onto a moving green bar. If the bar is visible only briefly, the flashed line appears yellow. If the trajectory of the green bar is exposed for sufficient time, however, the line is incorrectly perceived to trail the bar, and appears red.*" Nijhawan interprets this result in terms of motion processing. He notes that this motion processing occurs in the cortex rather than the retina in primates and he suggests that the motion cues in this antecedent mechanism are affecting the perception established at the retinal level. As Nijhawan describes it: "*the ability of motion cues to affect the perception of colour is consistent with the Young−Helmholtz−Maxwell notion of a 'central synthesis' of yellow."* We would differ with this interpretation since, as we pointed out above, regardless of what happens in any subsequent processing, in a three-cone model of color vision, the information on the actual spectral composition of an illuminant can not return once it has been discarded in the very first steps in visual perception.

The Ives result on the dynamic breakdown of metameric matches is so important and so fundamental to the understanding of the mechanism of human color vision that we had to investigate it further and try to repeat it for ourselves. I describe the details of this investigation in Chapter 6. In short, we did indeed observe the same disparity in the behavior of pure and mixed yellow under dynamic conditions as described by Ives. In our technique, we were also able to take these results a step further and observe separately, the perception by the two receptor classes (cones and rods). More specifically, we were able to simultaneously observe and distinguish the separate perception by the rod and cone receptors, which to our knowledge has never been directly and knowingly accomplished before.

I describe a set of experiments validating this claim by correlating the observed rod and cone perception with receptor distribution in the retina and through the observation and measurements of distinct patterns of dark adaptation of the rods and cones. We further made use of the demonstrable independence of wavelength for the time course of the rod perception as a fixed reference to measure the differential latency of cone vision as a function of the wavelength of a light source. In those measurements we unequivocally observed the shortest latencies for the longest wavelengths with latency increasing monotonically with decreasing wavelength. It is exactly this ordered latency that explains the breakdown of metameric matches under dynamic presentation conditions.

In summary, then, direct identification of *multiple cone types*, despite prodigious efforts in many laboratories, have remained elusive.

Single cone microspectrophotometry has not resolved the situation, and even if one had a complete specification of three cone pigments – through molecular genetics, for example – this would not fully explain the characteristics of human color vision (Sheppard, 1968). The actual evidence is that the wavelength composition of light is resolved at the detection level and that the limited dimensionality of the perception is related to the number of output channels conveying the information out of the cones in a characteristic temporal color code.

## The Cone Spectrometer Model

The lines of evidence outlined above indicate that individual cone color receptors can each resolve colors to some extent, as a kind of miniature spectrometer. There has traditionally been a good deal of resistance to this concept since no model of this type has yet been proposed which is both physically plausible in terms of the structure actually present in the human retina and also contributes to the understanding of human color perception.

It is the purpose of this volume to show that there is a previously unrecognized and viable alternative to the multiple receptor model, a conceptually simple scheme whereby the cone color receptors can function as a type of spectrometer in a manner consistent with both the retinal structure and the observed characteristics of human color perception. Multiple cone pigments are not required but neither are they precluded by this concept. Indeed, the existence of multiple pigments can potentially enhance and optimize the mechanism proposed here.

In the most elementary terms, the basic concept of the model is that in a tapered optical fiber which is of sufficiently small diameter (as scaled by the optical wavelength) the light incident at the wider entrance end is barely able to "fit" into and illuminate the cone interior; as light propagates down the length of the contracting fiber, it encounters even smaller diameter regions into which it can "fit" with progressively greater difficulty  The light intensity illuminating the interior of such an optical fiber is consequently attenuated along the cone length as the optical power shifts to the exterior region. This attenuation is differential with wavelength, the longest wavelengths (for which the scale size of the cone is smallest) attenuating the most rapidly. Thus, as a result of the physical diameter change, an optical fiber can physically sort light into a positionally correlated sequence. Light of all wavelengths can illuminate the wide entrance end of the cone and only the shortest wavelengths can illuminate the narrow, distal end of the cone.

This color separation effect is schematically illustrated in Figure 13. A cone is shown with white light incident at its wide entrance end. Initially, the white light - consisting, for example, of red, green, and blue (R+G+B) components – fully illuminates the wide diameter end of the cone. As it propagates down the length of a properly "tuned" cone the longer wavelength red light is excluded from the cone interior first, leaving only green and blue. At smaller diameters, the green is also excluded from the cone interior leaving only blue light to illuminate the cone. The diagram shows the cone partitioned into three separate regions of illumination consisting of R+G+B, G+B, and then only B. However, in an actual cone, the separation would be continuous with wavelength and would not occur in three distinct steps as illustrated. We also do not mean to imply that the cone is somehow divided into three

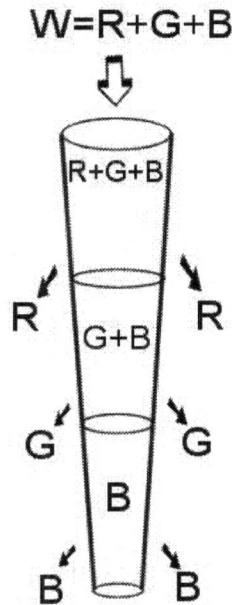

**Figure 13. Schematic representation of light illuminating a cone.**

separate regions where these particular color combinations are somehow detected and subsequently signaled out for further processing. While some sort of partitioning like this could actually work, I would suggest that the actual process for extracting the color information in retinal cones occurs through a much simpler and less cumbersome process.

A careful examination of this very simple color dispersion concept leads to some interesting conclusions. First, an analysis of the optical physics involved will show that not only is this color sorting mechanism a possible one in principle, but that its most basic features are in accord with the corresponding properties of human color vision. These include directional color shifts (Stiles-Crawford effect of the second kind) and other limitations to the kinds of color discriminations possible (such the perceptual similarity of violet and purple, and the existence of the special status of blue light)

Secondly, the possibility that this process is at work in the human eye is demonstrably supported by the available evidence on the physical structure of the retina and its photoreceptors. The retinal cones are of the right size and shape to exhibit the proposed wavelength-sorting process (including the foveal cones). The observed variations in cone structure across the retina account for the observed variations in color resolution in the corresponding retinal areas (in addition to the functional zoning of peripheral color vision mentioned previously).

That color information is thus potentially available in the cones and even that the characteristics of the information are similar to that of the perceptual function, is not alone sufficient to dictate that the proposed cone spectrometer model is the basic mechanism of human color discrimination One must, after all, have some assurance that the potentially available color information could be detected and utilized by the neural circuitry of the retina in a plausible way.

I will describe how that very structural organization of the retinal cones that enables them to spatially code color information in the first place, results in the automatic encoding of that information

40

through the involuntary saccadic eye movements in a characteristic temporal code. In Chapter 6 we will explore how a naturally occurring temporal coding scheme would operate and how it leads to a comprehensive accounting of subjective color phenomena such as the long-enigmatic Benham's Top. It will be seen that this model can lead in a very natural way to the opponent-color and trichromatic aspects of normal color vision. In addition, we will address how it is possible - through simple defects or mismatches that may occur in the color dispersion mechanism of the cones - that the various forms of color defective vision can come about.

Finally, the proposed model addresses what is arguably the most important unasked question about the color receptors of the eye: Why are the cone color receptors cones? That is, perhaps for the first time, this model addresses the simple fact that those receptors that mediate color vision are universally conical in shape since it is that very cone shape itself that plays the critical role in its color discrimination function. The only other proposal that I am aware of for a functional role of the cone shape is that of Miller and Snyder (1972) suggesting that peripheral cones are tapered to allow them to radiate light into the adjacent rod outer segments. However, this does not explain the cone taper in the central retina where there are few rods surrounding the cones although, in the more peripheral retina, this mechanism could conceivably play a role in enhancing the scotopic sensitivity of the eye (although this role has been discounted by some, c.f., Rowe, Corless, Engheta, and Pugh, 1996).

## But, What About Blue Cones?

At this point, a knowledgeable reader may be asking: "But what about blue cones?" Indeed, this is a very good question. It would seem that there exists a separate population of cones that have been interpreted as being the so-called blue or S cones. In particular, the seminal paper by Curcio, Allen, Sloan, Lerea, Hurley, Klock, and Milam (1991) appears to have definitively characterized a subpopulation of cones that have been identified as these blue cones. In their words, they "stained 7 wholemounted human retinas obtained from 6 female donors, using an affinity purified antibody to a 19 amino acid peptide sequence at the N-terminus of blue opsin, standard PAP immunocytochemistry, and controls". That is, they stained their retinal material with an antibody specific to the blue-sensitive opsin. Their staining technique labeled a subset in which they found that "foveal blue cones are sparse, irregularly spaced, and missing in a zone about 100 microns (0.35 degrees) in diameter near the site of peak cone density".

Their summary abstract states: "These findings are consistent with psychophysical reports of foveal tritanopia and maximum sensitivity to blue light at 1 degree eccentricity". In their introduction, in discussing related studies as well, they further state: "Many of these methods are inferential, and cells are presumed to be blue cones by virtue of the similarity of their distribution to that suggested by visual psychophysics". In short, they inferred that they had identified a subpopulation of cones stained with anti-blue opsin that fit the profile expected on the basis of the standard model. Clearly, Curcio, et al. (1991) identified cones that were different, but the nature of that difference is less than unequivocal. Apparently, the stained cones contained blue opsin, but the sparseness of the identified population (less than 7% of cones within 4 mm of the foveal center) raises the question as to what certainty there may be that other cones might not also stain with the anti-blue opsin under other circumstances.

They did note that the <u>inner segments</u> of the stained cones appeared to be morphologically different than that of other cones, being longer ("10% taller") and more cylindrical in shape. Perhaps morphological differences indicate that this cone subpopulation is more related to some entirely different process than solely color discrimination. For example, could these be cones that were undergoing some kind of photomechanical process such as elongating under dark conditions as have been noted for other species (although not for humans)?

Another possibility for these blue cones is raised by some observed similarities between these putative blue cones and rods. In addition to noting the similarity in distribution within the retina of their blue cones and rods, Curcio, et al. (1991) noted:

> "Recent studies have revealed additional similarities between blue cones and rods on a cellular and molecular level. For example, the amino acid sequence of blue opsin more closely resembles that of rhodopsin than does the red or green opsin (Nathans et al., 1986). Blue cones and rods both contain S-antigen (Müller et al. 1989) and lack carbonic anhydrase activity (Nork et al., 1990), and their phosphodiesterase inhibitors are highly similar (Hamilton and Hurley, 1990). Despite these similarities, however, the electrical response of blue cones is similar to that of R/G cones (Schnapf et al., 1990). Other, previously described parallels between the rod and blue cone system, such as poor spatial and temporal resolution (Zrenner, 1983) are probably explained by different mechanisms for the two cell types."

These observations raise the prospect of some intriguingly different possibilities. What if these so-called blue cones are indeed a different population of cones, ones that constitute a hybrid between cones and rods? These receptors have evident similarities and differences from both the major population of cones as well as the rods. They function like cones but have many rod-like characteristics. Perhaps they are an evolutionary approach to provide some color vision function at somewhat lower light levels or are perhaps a remnant of some evolutionary process bridging the transition between rods and cones. An interesting point in this regard is underscored by the observations of Xiao and Hendrickson (2000) who looked at the spatial and temporal expression of opsins in human fetal cones. They found that: "S opsin protein appears in and around the fovea at fetal week (Fwk) 10.9, whereas L/M opsin first appears in the fovea at Fwk 14-15. S opsin mRNA and protein are consistently detected much further into peripheral retina than L/M opsin, indicating that S appears before L/M opsin." That is, their observations clearly suggest developmental differences in what might be the blue cones as compared to all other cones.

It should be noted that Xiao and Hendrickson (2000) made some additional, very interesting observations. In their words: "Cones containing both S and L/M opsin (S+L/M) appear around the fovea shortly after L/M opsin is expressed, are found in more peripheral retina at older ages, and decrease in number after birth. Some S+L/M cones are still detected in adult retina." They go on to note: "However, the presence of cones containing both S and L/M opsin during development suggests that human cones can respond to the factors controlling expression of each opsin." Thus they too have found that individual cones of the human retina can express multiple pigments, a result that makes little sense in the three-cone model of color vision. Moreover, that both the S opsin and the longer wavelength absorbing opsins can be found in the same cone raises the question of just how unique were the blue cones observed in the study of Curcio, et al. (1991). That is, they apparently observed a separate population of cones containing the S opsin, but how many of these same cones may also have contained the L/M opsins? This question may need to be addressed in further studies since it clouds the issue the uniqueness of these blue cones despite the apparent morphological differences in their inner segments..

In any event, the outer segments of these blue cones have form and dimensions very much like the larger population of cones and, in terms of the cone spectrometer mechanism discussed here, should function much like any other cone – although perhaps with different efficiency at different luminance levels. Clearly, resolving the nature and function of these blue cones will require substantially more research and investigation although, perhaps, with a somewhat different approach than has been employed previously.

The remainder of this volume will focus on the several lines of evidence supporting the cone spectrometer model. We will explore the related aspects of how it must function and how it consequently explains many of the characteristics of human color vision function including many that are still unexplained or poorly understood.

# Chapter 3    Optical Transmission in Cones

The structure of the photosensitive layer of the human retina, consisting of a closely packed parallel array of intermixed rod and cone receptors has been likened to a fiber optics bundle    (Enoch, 1967).  In their essential features the receptors are aptly named; the photosensitive outer segments of the rods, generally believed to only mediate the achromatic vision at low (scotopic) light levels, are cylindrical in shape and those of the cones, generally believed to only mediate the chromatic vision at high (photopic) light levels, are cone-like in shape.

That the receptors are more refractile (have a higher refractive index) than their surrounding medium and that light can thus be confined to the receptors through total internal reflection was first noted by Brücke (1843).  The directional sensitivity of the retina, the Stiles-Crawford effect of the first kind (Stiles and Crawford, 1933), is generally understood in terms of the angular light acceptance properties of an optical fiber (Kapany, 1967) as implicated by Wright and Nelson (1936) and O'Brien (1946). Toraldo di Francia (1949) explicitly noted the small size of the photoreceptors, with diameters on the same order of dimension as the wavelength of visible light, and pointed out that the simple ray picture of light transmission of geometrical optics is not adequate to describe the attendant wave diffraction and interference effects.  The light transmission characteristics of small optical fibers are described by the wave optical theory of circular dielectric waveguides (Snitzer, 1961; Kapany and Burke, 1972; Marcuse, 1974; Snyder and Love, 1983); there it is seen that light is guided in characteristic waveguide modes (see the Appendix for the mathematical details of the waveguide solution for a circular dielectric waveguide).

The direct observation of light transmission in the form of these low-order dielectric waveguide modes in the receptors in retinal preparations (including that of the human retina) by Enoch (1961a) has left open the question of the role of such waveguiding behavior

45

in the visual process. There have been previous attempts to implicate the waveguide properties of the receptors in models of color vision. Barer (1957) was perhaps the first to recognize that the optical properties of the receptors could conceivably mediate color vision. He pointed out that transmission differences for different diameter fibers could, in principle, result in cones of different size having different spectral sensitivities even if they all contained the same photopigment. Myers (1962) suggested a specific trichromatic model of this type, based on three cone classes of different characteristic diameters. Biernson and Snyder (1968) suggested that each cone could resolve full spectral information by detecting which waveguide modes were transmitted in the receptors. For their model it was necessary to assume the existence of a complicated and rather unlikely radial scanning mechanism in order to detect the actual mode transmitted by the cone in a given instance. Neither of these models is particularly in accord with the actual structure of the retina (different classes of cone diameters in a given region of the retina are not observed and the best evidence indicates that the cones are too small to exhibit the multiplicity of modes required by the Biernson and Snyder model - see below) nor do they lead to an obvious accounting of the color vision phenomena. They are, however, indicative of the possibilities inherent in the waveguide concept.

## Light Propagation in Small Optical Fibers

In geometric optics light is represented by straight-line rays propagating in a direction normal to the wave fronts of constant phase of the radiation field. This description does not inherently contain the wave phase information, and for this reason fails when the scattering geometry is on the same order of dimension as the wavelength of light where that phase information critically determines the details of wave interference effects. There are, however, procedures available by which the phase information can be put into this formulation and the physical simplicity of the ray picture retained. This involves explicit computation of the phase shifts resulting from the radiation fields propagating along the optical path length of the refractive media and the contributions from each reflection at discontinuities in refractive index. (This procedure can be exploited only in cases of very simple geometry such as the slab waveguide. There, the results obtained are the same as for the wave optical picture as Marcuse (1974) explicitly shows. While the method is rather difficult to apply to the more complex geometry of a circular guide, the physical concepts developed for the simpler case are nonetheless instructive.)

In this formulation of the ray picture, the condition for light guiding, that is, confinement of the rays to the vicinity of a given region through the process of total internal reflection, is that the sum of the contributions to the phase shift (accumulating after each reflection) do not distort the fronts of constant phase of the radiation field. This obtains when the total phase shift of the waves between reflections adds up to an integral multiple of $2\pi$. This condition can only be satisfied for specific propagation angles of the rays with respect to the guide walls; rays incident at other than these specific angles will not be confined or guided in that region. When the width of the guiding region is large, many propagating angles are possible, so that we simply have the geometric optics limit where the phase information is not important and there is essentially a continuum of possible ray directions (up to the value of the critical angle) for total internal reflection.

In a circular fiber, all the "rays" propagating at one of the specific allowed angles with respect to the guide axis at all the possible azimuthal orientations about that axis will superimpose and their interference pattern will form a characteristic cross sectional distribution pattern (or standing wave) of the radiation field amplitudes. Each such pattern associated with each characteristic angle is a waveguide mode of the wave optical formulation. The standard computational procedure in the more rigorous wave optical formulation consists of solving Maxwell's equations for the electromagnetic fields and applying the appropriate boundary conditions for the particular geometry of interest (Kapany and Burke, 1972; Marcuse, 1974). In practice, this direct method of solution can be carried out only for simple geometries with tractable boundary conditions, such as the dielectric slab and the uniform circular cylinder. This method of solution leads to a complicated transcendental eigenvalue equation which must, in general, be solved by numerical computation methods.

Each eigenvalue solution of the boundary-value problem corresponds to a characteristic propagation mode of the waveguide. For the uniform circular fiber the properties of the waveguide solutions are determined predominantly by the dimensionless characteristic waveguide parameter V, given by

$$V = \frac{\pi \cdot d}{\lambda} \sqrt{(n_1^2 - n_2^2)} \tag{1}$$

where d is the fiber diameter, $\lambda$ is the free-space wavelength of the incident radiation, and $n_1$ and $n_2$ are the refractive indices of the fiber core and its surround, respectively (Snitzer, 1961). The refractive index difference, $\sqrt{(n_1^2 - n_2^2)}$ is just the numerical aperture, NA, of the fiber defining the maximum angle, $\theta$, of the acceptable cone of light rays that will be guided by the fiber through $NA = n_3 \sin\theta$, where $n_3$ is the refractive index of the medium in the region preceding the fiber as depicted in Figure 14 (Kapany, 1967).

The parameter V is just the product of the circumference of the circular fiber (in units of the radiation wavelength) and the fiber numerical aperture

$$V = \frac{\pi \cdot d}{\lambda} NA. \tag{2}$$

V is the relevant measure of the "size" of the waveguide; the larger V is (through either d or NA being larger or $\lambda$ smaller), the larger is the guide and the less important are waveguide effects and the more accurate will be the geometric optics ray picture. The smaller it is, the more important are the characteristic waveguide effects.

For guided modes, not all of the power in the radiation field is confined to the fiber interior. In terms of the ray picture, the reflections at the fiber-surround-interface entail excursions into the surrounding medium. These excursions (which are very small when the fiber is large) result in the so-called Goos-Hänchen shift and are the basis of internal reflection spectroscopy techniques (cf. Harrick, 1967) in which these excursions are used to probe the nature of the external medium. As the fiber diameter decreases (strictly speaking as V decreases) these excursions increase in amplitude as it becomes progressively more difficult to confine the light rays to the vicinity of the guide. At some point the guide becomes small enough that the rays propagating at a given angle can no longer be confined to the vicinity of the guide axis and the light radiates away.

This is the phenomenon of mode cut-off. In the waveguide terminology, the excursions into the external medium are the evanescent surface wave (exponentially damped outside the fiber wall) and part of the mode power is guided within the fiber and part in the external medium. The modal efficiency, $\eta$, is defined as the fraction of the

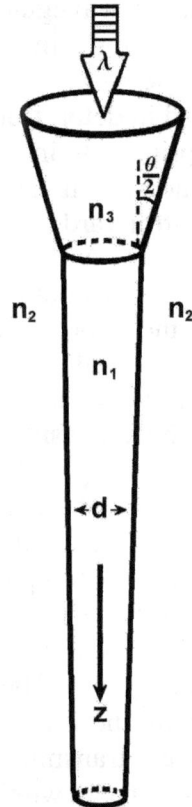

**Figure 14.** Idealized light guide geometry for an optical fiber of diameter d and refractive index $n_1$ imbedded in a medium of index $n_2$. Light of free-space wavelength $\lambda$ is incident along z-direction parallel to fiber axis.

49

total power guided by a mode which is actually propagated within the confines of the fiber wall. As V decreases, the modal efficiency drops to zero at a characteristic value for each mode.

The mathematical details of waveguide transmission in a dielectric fiber are derived in detail from first principals in the Appendix. It is a straightforward matter to write a computer program to solve the appropriate transcendental eigenvalue equations from this analysis for the possible modes. The results of such a numerical calculation are used to compute the cut-off curves for the efficiency of an optical fiber, depicting the fraction of the transmitted power carried within the fiber as a function of V, for the two lowest-order modes, the $HE_{11}$ and $HE_{21}$, and the result is shown in Figure 15. The cut-off curve for the $HE_{21}$ mode is typical of all

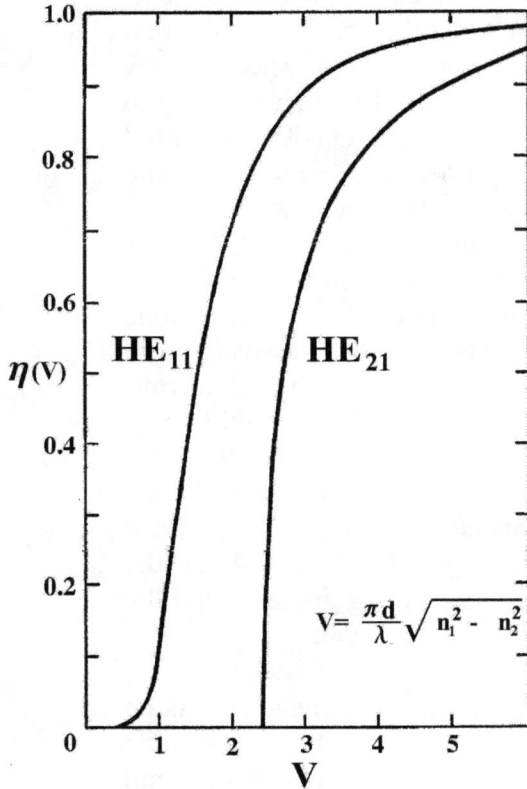

Figure 15. Cut off curves for the lowest-order cylindrical waveguide modes. The modal efficiency, $\eta$, of a dielectric circular fiber is plotted as a function of the dimensionless wave-guide parameter V. Optical power can propagate within a fiber with V < 2.405 only in the $HE_{11}$ mode.

The figure shows $\eta(V)$ on the vertical axis (0 to 1.0) versus V on the horizontal axis (0 to 5), with curves labeled $HE_{11}$ and $HE_{21}$, and the relation:

$$V = \frac{\pi d}{\lambda} \sqrt{n_1^2 - n_2^2}$$

the higher order modes where the modal efficiency drops abruptly to zero at a critical value of the waveguide size and the energy is radiated away. For V = 2.405 or less only the $HE_{11}$ mode is not cut off. The $HE_{11}$ mode does not exactly cut off except for V = 0. However, very little power is carried within the fiber for small but non-zero values of V. For example, at V = 0.3 less than one part in $10^{16}$ of the incident electromagnetic energy is within the guide and the mode is effectively cut off for all practical purposes. The radial distribution of light inside and outside a fiber at various values of the waveguide parameter are explicitly computed and shown in the Appendix for the $HE_{11}$ (Figure 49) and $HE_{21}$ (Figure 50) modes.

In terms of the photoreceptors of the human eye, these concepts are relevant to visual function, since it is only the power of the radiation field actually propagated within the outer segment of the receptors that can be absorbed by the photolabile pigment resident there. The receptor transmission characteristics (which are wavelength dependent) determine the intensity of light illuminating the receptor interior. Thus, the photon energy absorbed by the receptors depends not only on the photopigment absorption spectrum and its concentration but also on the physical geometry of the receptors.

Of course, since the above description of dielectric waveguide phenomena is derived for an ideal, uniform, and infinitely long cylindrical fiber, there are some important limitations to the accuracy with which it can represent light transmission in real retinal receptors. In the first place, the physical parameters of the receptors are difficult to measure accurately -- thus leaving one with some uncertainty as to exactly what values of parameters to model. In the second place, as real structures, the receptors can be expected to have local irregularities of form or inhomogeneities of refractive index, nor are they infinitely long, having a beginning and an end.

In the most general of terms, the relevant measure of the departure from ideality is the scale size set by the wavelength of light. Thus, if the local irregularities of receptor construction are small compared to the wavelength of light, the passage of light through the receptor will be little perturbed and there will be little scattering. The lamella arraignment of the outer segment membranes, for example, can be treated as a homogeneous medium as far transmission is concerned since the membrane repeating distance is an order of magnitude smaller then the wavelength of light.

That the receptors are not infinitely long does not negate the usefulness of the ideal description. While the conditions determining the details of how the electromagnetic fields are coupled into the cylinder are complex, at distances of only a few wavelengths from the ends, the infinite rod description is a good approximation.

The light guided within the photosensitive outer segments of the receptors is subject to being absorbed by the photopigment, and the incident radiation is exponentially attenuated thereby. The absorption is mathematically represented by an imaginary component of the refractive index (the extinction coefficient) which leads to exponential damping of the electromagnetic fields (Pankove, 1971; Fischer and Rohler, 1974). Again, however, if this absorption is small over distances comparable to the wavelength of light, the mode transmission characteristics of the fiber are substantially unaltered (Snyder and Pask, 1973; Marcuse, 1974). In fact, measurements (at least in goldfish cones) show that the imaginary part of the refractive index is trivial compared to the real part (Hárosi and MacNichol, 1974; Rowe, Corless, Engheta, and Pugh, 1996). The absorption itself is rather important to visual function, of course, since it is only photons absorbed by the pigments which can contribute to a physiological effect. The absorptive nature of the receptors also leads to anomalous dispersion of their refractive index (Snyder and Richmond, 1972), although the influence of this anomalous dispersion on receptor spectral sensitivity is expected to be small (Stavenga and van Barneveld, 1975).

The primary focus of the present study is the effect of the conical shape on the optical transmission properties of the retinal cone outer segments. The general problem of the non-uniform tapered waveguide is, of course, more complex than the uniform cylinder.

One possible approach is to approximate the tapered section by a uniform cylinder with an effective diameter characteristic of the effect of passing through the conical portion (Snyder and Pask, 1973). Such a description does, of course, subsume the details of what happens within the length of the tapered section and can only tell what is the overall result of passing through the cone (total absorption, for example).

The waveguide color vision schemes mentioned previously (Myers, 1962; Biernson and Snyder, 1968) describe light transmission in the retinal cones by representing the cones as uniform cylinders. At first sight, this might seem sufficient since the foveal cones of the human retina, exhibiting the most highly developed color vision, are after all only slightly tapered and rather rod-like in appearance. However, the presence of even a small residual taper in the foveal cone outer segments (see below) instead of an exact cylindrical shape can be very important if one would like to identify a spectroscopic principle and wants to know what is happening to the transmitted light within the cones.

A more detailed theoretical approach is to represent the tapered guide as a succession of separate cylinders, each of infinitesimal length and each of successively smaller diameter. The diameter of each successive cylinder is equal to the actual cone diameter at each point. One can then depict the cone solutions in terms of what Snyder (1970) has termed the "tapered" or "local" modes; at each point along the tapered guide, the solution is locally in terms of the modes of the uniform cylinder (cf. Marcuse, 1974; Snyder and Love, 1983).

The solutions of the tapered guide are thus combinations of the cylinder modes and this representation can be accurate since all the modes of the uniform cylinder (including the discrete set of bound modes both forward [transmitted] and backwards [reflected] as well as the continuous set of unbound [radiation] modes) constitute a complete orthogonal basis set, suitable for the representation of an arbitrary function. The greater the departure of the tapered segment from the cylinder shape, the more terms of the basis set which must be included to accurately represent the tapered modes. The expansion coefficients weighting the contribution of each cylinder mode to the basis representation changes as a function of position along the cone axis (z-axis).

With this computational approach, Snyder (1970, 1971) finds that the cone taper induces coupling between the waveguide modes, so that an initially pure single mode entering a tapered section will have part of its power transferred to the other cylinder modes. As one might expect, the importance of this power transfer is measured by the rate at which the cone diameter changes over distances comparable to the wavelength of light. Power transfer is small if the tapering is small. The power transfer is proportional to $(\tan^2\theta)/(\beta_1 - \beta_2)^4$ where $\theta$ is the cone taper angle and $\beta_1$ is the propagation constant for the initial mode and $\beta_2$ that for the mode to which the coupling is computed. This coupling is quite small for small taper angles and is increasingly negligible with increasing difference in propagation constants. Only the nearest neighbor modes, for which the difference in propagation constants is smallest, need be computed and this coupling is only significant where the taper angle is large (Snyder, 1970; see also Bures and Ghosh, 1999).

Of particular interest is the case where the local value of waveguide parameter V of the cone is less than 2.405; then the only discrete bound mode which can be supported by the fiber is the lowest order $HE_{11}$ mode and we may neglect coupling to the other discrete modes (Snyder and Love, 1983). While for small enough guides there will thus not be any coupling to the other discrete forward modes (out of the $HE_{11}$ mode) one must also consider coupling to the other kinds of waveguide modes (backwards modes and radiation modes) as well.

The backwards modes are identical to the forward bound modes except that they are propagating in the reverse direction in the waveguide and have the same propagation constants as the corresponding forward modes except with the reversed sign. The negative sign on their propagation constants means that the coupling to these modes will be very small indeed when compared to that for the forward directed modes (the denominator in the above expression for the coupling is then anti-resonant).

Marcuse, 1970, has computed the coupling to the continuous modes of the radiation field. The power losses are found to be small, especially for slightly tapering sections. Marcuse confirmed the predictions using Teflon cone models and appropriately scaled microwave radiation. For the case where a tapered section joins two cylindrical regions differing in diameter by a factor of 2.0 in a waveguide with $V \approx 2.5$, even for the worst case situation where the length of the tapered section is zero and the diameter change is abrupt, the power losses to the radiation field are only 22% of the incident power. For a connecting section with a taper angle of $10°$, losses are about 14% and for a taper angle of $0.5°$ power losses are less than 1%.

Thus the net result of all these considerations is that for sufficiently small ($V \leq 2.405$) and sufficiently gently tapering (say $\theta < 1.0°$) the transmission conditions for light guiding in a cone are rather uncomplicated: light is bound to the guide in the $HE_{11}$ mode and little power transfer to any of the other modes of the waveguide occurs. Of course, the waveguide parameter V is decreasing along the length of the cone as the diameter decreases and thus the $HE_{11}$ modal efficiency is decreasing or cutting off along the length of the cone. The result is that the optical power is gradually shifted out of the cone interior into the evanescent surface wave outside the guide. Simply stated, the intensity with which the cone interior is illuminated decreases with decreasing fiber diameter. The most significant point in the present context is that this attenuation is differential with wavelength. The waveguide size, V, is smaller for larger wavelengths so that these attenuate more rapidly along the cone than do the shorter wavelengths (Medeiros, Borwein. and McGowan, 1977).

How the light intensity illuminating the cone interior attenuates along the contracting taper is graphically displayed in Figure 16. The cone model used for the computations is an idealized version of a foveal cone outer segment (see below): 0.8 μ wide at its entrance end, tapering over its 40 μ length with a full angle of less than half a degree (26'). For this plot, NA = 0.22 and thus V = 0.7 d/λ. The logarithm of the relative waveguide efficiency (which is the same as the logarithm of the relative light intensity within the cone) is plotted as a function of position along the cone for 50 nm wavelength intervals spanning the visible spectrum. As is apparent, the differential attenuation of light of different wavelengths can be a large effect. In this cone model, red light of 650 nm wavelength attenuates more than two orders of magnitude faster over the cone length than does 450 nm blue light.

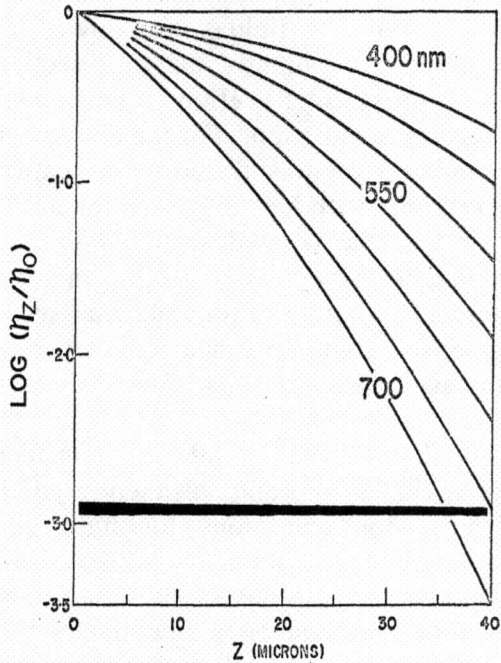

Figure 16. Wavelength-differential attenuation of light in a dielectric cone of geometry comparable to human foveal cones (indicated in diagram). The logarithm of the transmission efficiency at each point along the guide relative to that at the entrance is plotted for a cone with (πNA) = 0.7 (this is the same as a plot of the log of the relative intensity remaining to illuminate the cone interior). In this case (0.8μ proximal diameter, 0.43° full taper angle and 40μ length) light at the long-wavelength end of the spectrum attenuates a thousand times faster than for the short-wavelength end.

# Direct Observation of Color Separation in Cones

We present here some direct observations of spectral dispersion along the length of an optical fiber. These were conducted with small (2 mm in diameter) rods of fused quartz ( n = 1.4585) that were heated in a zone around their midsection and allowed to be drawn out into a taper as the weight of the lower half was allowed to taper the fiber. One of the resulting sections was then mounted within a cell (with a glass window on one side) that was filled with a ~66% sucrose solution with a slightly lower refractive index that nearly matched that of the rod. A white light source was focused onto the broader entrance end of the glass rod and the tapered section was observed from the side of (and slightly below) the rod.

The resulting transmission was then observed and photographed with magnifying optics. The original color photographs of the resulting spectral dispersion along the tapered rod are shown on the cover of this volume. Reproduced here in Figure 17 is a black and white version (because of the publication limitations of this book) of the tapered rod within the cell. This shows the overall geometry of the set up and the large light losses out of the rod along the tapered section (before the small narrow tip of the fiber that is shown separately below).

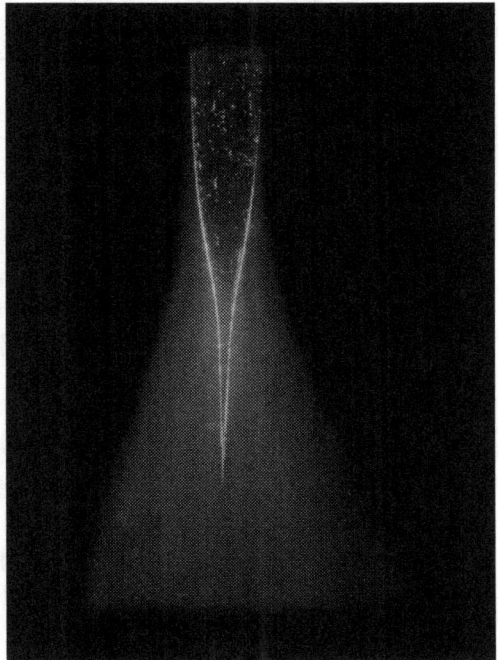

Figure 17. Tapered glass rod in a near matching refractive index liquid showing light leakage along the taper. Spectral dispersion is only observed near the end of the small tip of the tapered rod.

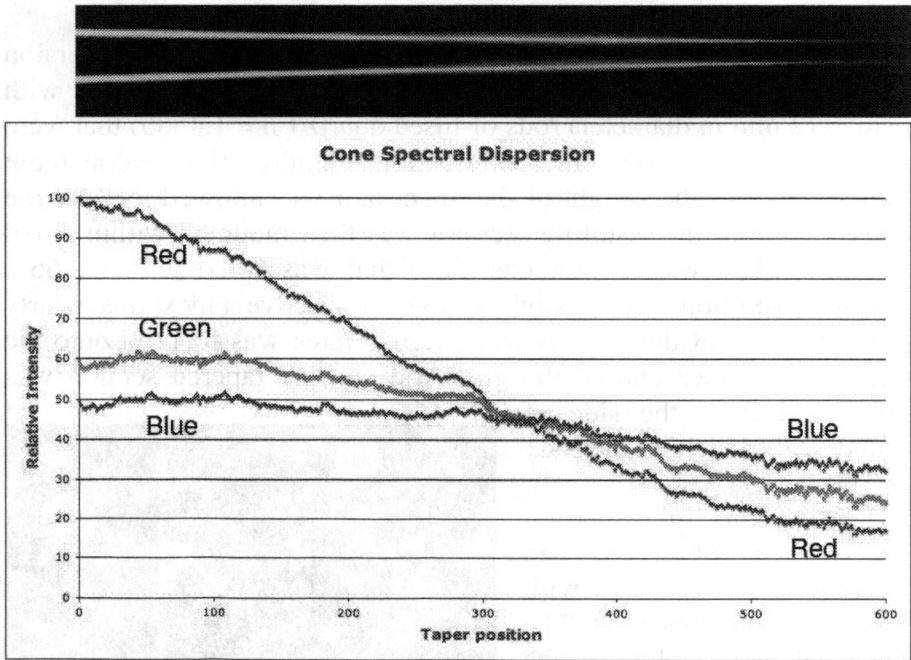

**Figure 18. Spectral dispersion along a tapered glass fiber.** As a result of the longest wavelengths attenuating the most rapidly, the data plots of red, green, and blue light intensity reverses from the ordering at the beginning to blue, green, and red at the end

The experiment is remarkably easy to perform. In our case, for the setup in Figure 17, we used a fluorescent dye (disodium fluorescein) in the liquid in order to visualize the large light leakage of the higher order modes out of the tapered rod. Figure 18 is a close up view of the dispersion along the small tapered section very near the fiber end where the rod diameter is on the order of 10 μm. The figure also includes a plot of the relative spectral components of the light as it emerges from the cone. Of course, a black and white image of the cone tip will have little impact since the spectral dispersion (evident in the comparable, color version of the photograph on the front cover of this volume) is not obvious (a black and white photograph is, after all, the equivalent of monochromatic vision!). Of direct interest here

58

are the corresponding plots of the attenuation of the red, green, and blue components along the taper. The key feature to note is that, just as predicted by the basic physics of the optical propagation, light of the longest wavelengths is shunted out of the cone first and the shortest wavelengths last, in the smallest diameter regions.

Quite clearly - irrespective of the obvious relevance to what might be happening in the retinal cones - this effect can form the basis for the construction of miniature spectrometers. As displayed here, one need only detect the light intensity being excluded from the cone interior as a function of position along the length of the taper to determine the corresponding wavelength. Since this will necessarily occur (in the optical domain) with very small fibers, it is possible to construct spectroscopic devices with correspondingly small dimensions.

The cone model discussed here is highly idealized, since the calculations were carried out without considering coupling effects at the entrance and exit ends of the cone. The effect of absorption by any photo-labile pigments within or around the cones was also ignored for these calculations. While both coupling and absorption must take place in order to detect the radiant energy distribution, these effects can be made arbitrarily small. By using a sufficiently smooth input coupling taper, efficient fiber termination, and low pigment density, one could have radiant energy distributions that differ little from these idealized plots.

A more fundamental limitation on the design and construction of "cone spectrometers" is that such a device does not establish a one-to-one correspondence between the illumination at a particular point along the cone and the wavelength of the light reaching that point. Unlike a conventional prism or grating spectrometer, the extraction of color information depends on resolving the entire pattern of illumination along the cone. Since light of all wavelengths must pass through the broad entrance end of the cone, a detection event there by itself conveys little direct information about the color of the incident light (except perhaps, its intensity). However, if light is detected within the cone at the entrance end but is not detected at the small diameter exit end, one can conclude that the incident light contains predominantly long-wavelength components. On the other hand, a detection event within the small diameter end of the cone does by itself does convey the information that the incident light contains short-wavelength components, since red light is excluded from reaching that region by mode cutoff. That is, only blue light can be present within the smallest diameter portion of a cone and detection of absorption events there signifies the presence of blue with a high degree of certainty. The overall resolution with which one can detect color information with a cone spectrometer is evidently a function of the resolution with which one detects the overall illumination pattern.

These considerations suggest that one way to build a cone spectrometer would be to form a small tapered fiber using core material containing a photolabile substance. The core material must transduce a light absorption event into an electrical signal. If the photoabsorption site acts as a localized source in such a cone, then the possibility exists of detecting the pattern of energy absorption. If the absorption site can be localized to one of two regions - the large diameter half and the small diameter half, for example, then dichromatic color information would be available. Localization to one of three regions would produce trichromatic color information. This localization of the absorption site could be arranged by physically dividing the cone into lengths with separate electrical outputs. Comparison of the signal from each region would then allow one to extract the spatially ordered color information.

Although this scheme is conceptually simple, such a device would likely be difficult to build. In particular, it would be difficult to electrically partition the cone length without affecting the optical transmission properties of the fiber. Perhaps a more elegant and less cumbersome method of extracting color information is to convert the spatially ordered pattern into a time-ordered pattern. This can be accomplished without physically partitioning the cone at all. If, for example, it takes longer for signals generated at the small, distal end of the cone to reach the cone electrical output than it does signals

60

generated at the larger, proximal (entrance) end, there will be a natural correlation between the rise (or fall) time of any signals generated and the color of the light that produced that signal.

If the optical input is modulated (by either chopping the incident light or scanning an imaged scene over the cone input), then the temporally ordered color information would be impressed on the cone output. This information can then be processed by analyzing the transient electrical signals so generated in order to extract information on the relative energy deposition patterns. Simple inter-comparison of the early half of the rise time (or fall time) of the cone's electrical output with the latter half would give dichromatic color information. Resolving the transient cone signals into three portions would similarly provide trichromatic color information.

## *Reversed Propagation*

In a very simple experiment, Brindley and Rushton (1959) provided a critical test of possible human color vision theories. They compared the perception of color for light incident at the retina from two opposite directions: in the normal physiological direction (forward incidence) and in the reverse direction by illuminating the retina through a glass rod in contact with the back scleral coating of the eye of an observer looking as far to one side as possible. They were able to thus probe color perception for light incident from opposite directions in a retinal region at about $30°$ in the periphery.

They found that the apparent color of the test light was essentially the same for the two different directions. This result clearly rules out any possibility that human color vision could be based on the presence of selective color filters situated <u>in front</u> of the receptors (a proposal which has never been taken very seriously in terms of human color vision, although the existence of colored oil droplets in the retinal cones in some species clearly does influence their color perception). The widely held interpretation that this result also rules out waveguide models for color vision is, however, not correct, as has previously been pointed out by Enoch (1963).

61

The point is that any waveguide model would be dependent on the transmission characteristics of the cones and the wavelength dependence of these characteristics is not different in the two cases of light transmitted in a forward or a backward mode. This is simply the fundamental physical principle of the reversibility of the light path in an optical system known as the reciprocity theorem (or reversion theorem) of Helmholtz (cf., Born and Wolf, 1959, p. 380 and also Choi, Doble, Lin, Christou, and Williams, 2005).

Clearly, the light incident in the backwards direction will not couple into the outer segment very efficiently; the light does not have the advantage of funneling through the inner segment in this case and consequently there will not be the normally-present impedance matching at the ellipsoid to couple the light into the small photosensitive outer segment (di Francia, 1949). Thus, we may expect that the overall illumination of the outer segment interior will be less for the same light intensity if incident backwards as compared to forwards. After correcting for the attenuation in light intensity due to its passage through the thickness of the scleral coat, Brindley and Rushton did indeed observe a decreased sensitivity for backward-incident light.

Nonetheless, in terms of the present model, it will still remain true that short-wavelength light would couple into the smallest regions of the cone outer segment more efficiently than the longer wavelengths. That is, the ordering of the color information (and the perception derived from it) will remain substantially unaffected for backwards incidence even though the overall intensity for all wavelengths is lower.

# Chapter 4     Retinal Cone Parameters

For the proposed cone spectrometer scheme to be a possible model for human color vision requires that the retinal cones be of the right size and shape to exhibit the proposed wavelength dispersive effect. Prediction of its optical transmission requires knowing both the cone outer segments' physical dimensions and their numerical aperture (relative refractive index). Given that the receptors are so small and fragile, the required measurements are rather difficult, and since the emphasis in previous models has, by and large, been on pigments rather than structure, much less has been done in the way of measurement of human or other primate cone structure than would be desirable for the present modeling. However, what little information is available does place the retinal cones' physical parameters in the right range for cone spectrometer operation.

An important observation in this regard is that the cone photoreceptors outer segments are substantially smaller than they need to be to just provide the measured visual acuity of the human eye. That is, the minimum blur circle of plane waves at the retina is substantially larger than the size of the diameter of the cone's photosensitive portion. It is only through the use of adaptive optics techniques that make use of wavefront sensing of light reflected by the retina that spot sizes smaller than the retinal cone's (larger) inner segment can be imaged on the eye (Hofer, et al. 2005). This certainly begs the question as to why the color receptors are as small as they are since on grounds of simply catching photons they could be substantially larger. Are the cone outer segments this small to provide their color dispersion properties?

## Structure and Dimensions

Much of the available information on primate cone structure comes from the early light microscopy observations. Figure 19, schematically displaying the dimensions of the human retinal cones and their outer segments over the extent of the retina, is based on the observations of von Greef, 1900 (drawings reproduced in Miller and Snyder, 1972) except for the foveal cones whose dimensions are based on more detailed considerations discussed below. While the cone shape changes considerably over the extent of the retina, most of the difference in shape is a result of changes in outer segment length as opposed to changes in its width. Note also that, while the inner segment diameters increase with increasing retinal eccentricity – presumably to enhance the overall cone sensitivity – the photosensitive outer segment diameters do not increase as dramatically.

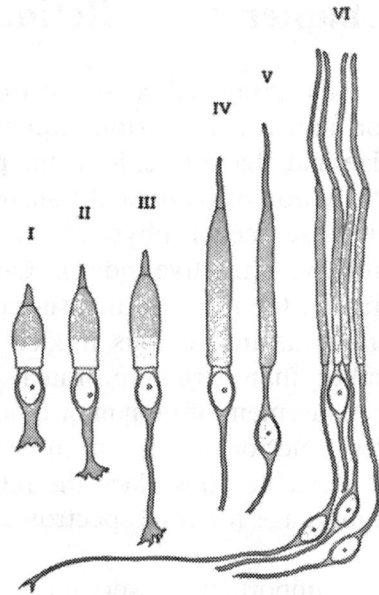

**Figure 19. Cone structure across the retina as revealed in the early studies of von Greef. Cones are portrayed in correct relative dimensions from the far periphery (I) to the fovea (VI)**

The diameter of the cone outer segment at its junction with the inner segment (its entrance end) increases from about 0.8 μ for foveal cones (Polyak, 1941; Dowling, 1965) to about 1.8 μ for peripheral cones. At the distal end, the diameter increases only slightly in going from the central to peripheral regions, from about 0.5 to 0.8 microns. The percentage decrease in diameter (and the corresponding decrease in cutoff parameters) over the length of the cones thus varies from about 40% centrally to 60% peripherally. The length of the cones decreases from up to 40 microns centrally to only 5 microns in the extreme periphery. The corresponding full cone taper angles thus increase from as little as 0.5° or less for the central cones to more than 10° for the peripheral cones. Most of this increase in taper angle is a result of a decrease in cone length.

The classical data on which the details of cone structure and its variation across the retina of Figure 19 are based, is somewhat approximate as to exact retinal position angle of the various cone "types" and there is not yet a modern comparable systematic examination of the retinal cone dimensions (but see Borwein, et. al, 1980 and Curcio, Sloan, Packer, Hendrickson, and Kalina, 1987). A synopsis of the receptor outer segment shape is shown with corresponding retinal location in Figure 20. (This is the same diagram displayed on this book's front cover except that there we have artificially colored-in the expected spectral illumination in the cone interior). The available data clearly displays a pattern of uniform and continuous alteration in cone shape across the human retina. Concomitantly, one also finds a systematic variation in color discrimination across the retina. While, as mentioned previously, color vision is organized in more or less well-defined zones

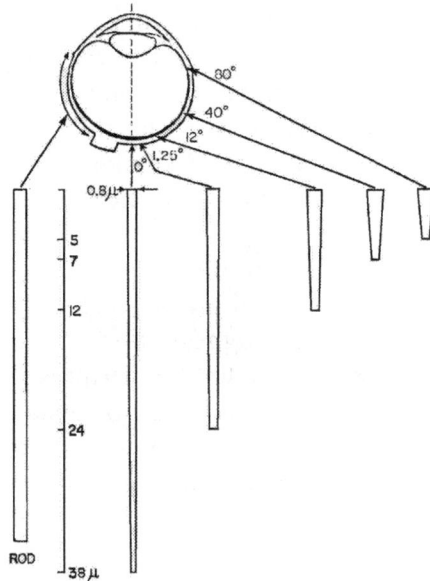

**Figure 20. Retinal receptor outer segment geometry across the retina as revealed by classical light microscopy. The primary change in cone structure for increasingly peripheral positions is a decrease in cone length, giving the cones a progressively more squat appearance. A typical rod outer segment is shown for comparison.**

of decreasing dimensionality out to the peripheral regions of the retina, the color sensations perceivable centrally are not entirely lost peripherally, since as Wooten and Wald (1973) have emphasized, by sufficiently increasing either the stimulus area or intensity the centrally-observed color perceptions can be recovered to some extent in the periphery. Thus color discrimination is not entirely lost peripherally, although it decreases markedly in resolution (in addition to its approximate zoning). Ideally, one would like to have data permitting one to plot color discrimination as a function of retinal position (for example, minimum discriminable wavelength differences systematically determined over the extent of the retina). Such a determination is made difficult by the rapid adaptation of the peripheral retina (whereby when a spot of light is incident on a given area of the retina, the retinal sensitivity there decreases and the apparent brightness of the light diminishes). This adaptation can be so rapid and marked in the peripheral retina that a test light can disappear altogether (Troxler's effect, cf. Clarke, 1960).

Despite these limitations, the available data on color discrimination in different retinal areas (Weale, 1953; Boynton, Schafer, and Neun, 1964) strongly suggests a systematically decreasing color resolution from the central to the peripheral retina (with the decrease being most rapid within a short distance of the foveal center). The proposed cone spectrometer model directly suggests a causal connection between the observed variation in cone shape and color resolution.

In this model the most important characteristic of the retinal cones enabling them to discriminate color is the size and shape of their photosensitive segment. If they are small enough that they are near critical cut-off for low order waveguide transmission and if their diameters decrease along the length of the segment, then different colors can be resolved as different patterns of photon energy deposited along the length of the cone. The resolution with which this pattern information can potentially be read is constrained by the length over which it is spread. That is, regardless of how the positionally correlated color information provided by a cone might be read, the resolution with which that information can be read will evidently be greater for encoding over a greater length. Thus in terms of variations in cone shape and color resolution over the extent

66

of the retina, the model provides an obvious explanation: the longer the cone, the better the color resolution it can mediate.

In view of the conceptual simplicity of the proposed cone spectrometer model, why has it not been previously proposed? (Actually Schroeder, 1960 had recognized that color discrimination was possible in principle by the cone dispersion of light as proposed herein. He dismissed the model at that time, however, on the basis of a misimpression about the "size" of the cones. He believed that the retinal cones were too large to effectively disperse the spectrum.). Probably the key impediment has been the widespread general impression that the "rod-like" foveal cone outer segments are actually cylindrical. The foveal cones do indeed have a "rod-like" appearance. If one is inclined to model human color vision in terms of multiple pigments, then the fact that the cones mediating the highest resolution are the least cone-like in shape might appear to warrant the dismissal of a role of the cone shape in receptor function. However, as indicated above, high resolution operation of a cone spectrometer requires spreading its diameter change over a relatively long length, i.e., we can expect a high resolution cone to look "rod-like".

Accurate measurement of the foveal cone outer segment physical dimensions is complicated by their relatively great length and small width and the rapid variation in these dimensions with position within the fovea. Polyak (1941, p. 448) summarizes the dimensions obtained through light microscopy by a number of investigators. The reports depict an outer segment nearly 40 $\mu$ long and less than 1$\mu$ wide. While some of the reports cite only a single number for the segment width, the studies which explicitly report measurements for both the base and tip of the outer segment give respective dimensions of approximately 0.8$\mu$ and 0.5$\mu$. This 38% diameter change over the segment length gives a full cone taper angle of 0.43°.

Since most of the emphasis in color vision modeling has traditionally been placed in photopigments rather than receptor structure, it is perhaps not too surprising that there has been rather little interest in determining the precise shape of the foveal cone outer segments and that some observers have simply reported cone diameters as a single number (even for cones which taper quite obviously). In some reports, observers have not even made a distinction between diameter measurements on the inner or outer segments of the cones.

No underline studies of primate cone shape have been reported in the modern era utilizing the improvements in resolution made possible by the electron microscope until 1980. In the one study using transverse electron microscopy on primate (rhesus) foveal cones (Dowling, 1965), the outer segments were reported as being 0.9 $\mu$ in diameter and not tapering. Dowling (1965) employed longitudinal thin sections through the receptor layer (parallel to the cone axis) for his observations. Given their great length and the expected small magnitude of the outer segment taper angle, the apparent shape of the foveal cones seen in thin longitudinal sections will depend critically on the relative orientation of the plane of sectioning and the cone axis. In a somewhat different context O'Brien (1951) has previously emphasized that serial transverse sectioning through the receptor layer is required for the determination of cone dimensions.

As Cohen (1972) has pointed out, while the foveal cone outer segments may not seem to taper significantly it is rather difficult to rule out the possibility of a slight taper on the basis of the EM observations. In terms of the proposed model the difference between a very slight taper and no taper at all is crucial. Strictly speaking, the cone "taper" does not necessarily have to be one of physical size change, since it is the change in the waveguide parameter, $V$, which will differentially attenuate spectral lights. Consequently, the "taper" can be one of decreasing numerical aperture without a dimensional change. However, there is no compelling indication of alterations in refractive index along the receptor length (which could result from changes in inter-disc spacing, for example), nor does it seem necessary to invoke such alterations, since a slight dimensional tapering of the foveal cones is clearly implied by both the overall intraretinal pattern of cone shape and as well as the direct observations in the classical light microscopy.

It was specifically to address these questions of foveal cone taper that we conducted a study of retinal cone geometry where we

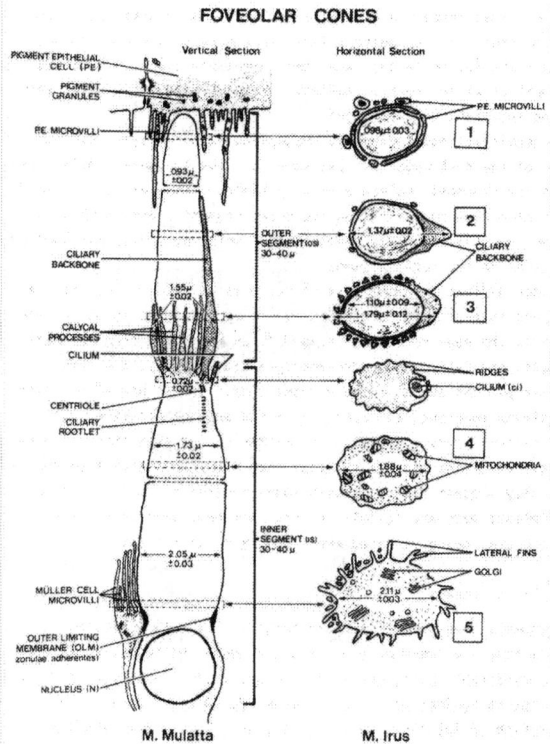

Figure 21. **Dimensions of the cones of the central fovea in monkey retina in transverse sectioning through the photoreceptor layer (from Borwein, Borwein, Medeiros and McGowan, 1980).**

69

specifically employed the technique of transverse sectioning (Borwein, et al., 1980). The retinas of three different monkey species were studied: Macaca mulatta, Macaca irus (fasicularis), and one Cebus monkey. In all cases, the retinal cones – specifically including the cones of the very central fovea (or foveola) – were found to indeed be tapered. The results for two species, including some of the observed structural details and measured cone diameters are shown as Figure 21.

## Refractive Indices in the Retina

Specification of the waveguide parameter for the retinal cones requires knowledge of the refractive indices of both the receptor ($n_1$) and its surround ($n_2$). While the state of experimental knowledge with regard to the refractive indices is hardly satisfactory, what is known is sufficient to unequivocally place the human cones very close to the right operating range and is consistent with their being in the ideal range for cone spectrometer operation.

Sidman (1957) has provided some of the only experimental measurements of primate cone refractive indices. He used an immersion medium index-matching technique on isolated photoreceptors from a number of different species. He found a remarkable uniformity in measured refractive indices in the corresponding portions of the receptor cells in the different species, despite considerable differences in the form and dimensions of their receptors. The best value of refractive index so determined for the cone outer segments is $n_1 = 1.387$ (c.f. Barer, 1957).

The refractive index of the interphotoreceptor matrix around the cones has not been directly measured, although Barer (1957) has given an estimate for its possible value. The refractive index certainly cannot be less than that of saline, 1.334. Indeed it must be larger than this value because of the suspended solids in the medium. Barer suggests a value comparable to that of serum, $n_2 = 1.347$. The interphotoreceptor matrix is known to be composed of a complex mixture of suspended mucopolysaccharides and proteins (Fine and

70

Zimmerman, 1963; Hall and Heller, 1969). It has been noted (Feeney, 1973a,b) that this matrix is significantly denser in the human retina than in those of other species, suggesting that Barer's estimate may be conservative. Even so, using $n_1 = 1.387$ and $n_2 = 1.347$ gives the numerical aperture as 0.33, so that in this case we have $V = 1.04 \, d/\lambda$. For light in the middle of the visible spectrum of 550 nm wavelength propagating in a 1.0 μ diameter region (larger than any part of the foveal cone outer segments) the waveguide parameter is less than 1.9. Thus, for these values of refractive index, the retinal cone waveguide parameter is below cutoff at visible wavelengths for all but the lowest order $HE_{11}$ mode and is sufficiently small to exhibit the kind of color dispersion mechanism we propose.

The accuracy of Sidman's refractive index measurements has been questioned since Sidman (1957) did not explicitly allow for possible post-mortem alterations and light induced damage in the handling of his preparations (Enoch and Glisman, 1966). But careful measurements of frog rods using an optical fringe-displacement technique reveal only minor changes in receptor dimensions and refractive index on bleaching (Enoch, Scandrett, and Toby, 1973) and -- in at least this case of frog rod outer segments -- confirms the accuracy of Sidman's measurement (1.400 by Enoch, et al., 1973, and 1.411 by Sidman, 1957).

Enoch's (1961a) observations of waveguide modes in human retinal preparations have been interpreted as requiring that the retinal receptor waveguide parameter be larger than implied above (Biernson and Snyder, 1968). The cutoff for the highest order waveguide mode Enoch observed required that V be at least 5.5 in some receptors. Because of the technical limitations imposed by the restricted depth of focus of the viewing optics and the great length of the foveal cones, Enoch found it necessary to observe extra-foveally. There, the cones are of larger width than in the fovea and rods are present. The waveguide size of the rods is rather larger than the cones: rods in the near periphery are 1.5 microns in diameter (Polyak, 1941). Thus, using $n_1 = 1.41$ and $n_2 = 1.347$ we would have

$V_{rods} = 3.42$ for $\lambda = 550$ nm (0.55 microns).

Enoch (1961a) subjected the retinal preparations to high intensity illumination prior to the observations (thereby bleaching the receptor pigment and negating any explanation of the wavelength dependent effects he observed in terms of differential pigment absorption). This too might affect the receptor waveguide size, depending on the amount of swelling or other disruption of the receptors thereby induced (Enoch and Glisman, 1966). While the alterations in waveguide parameter produced by bleaching might be small (Enoch, et al., 1973) the effect of mounting the retinal sample in saline as Enoch did might be significant. Although Enoch (1961a) did note that upon subsequent histological examination the "ground substance" between the receptors appeared to remain intact, the influence of the immersion medium refractive index on his retinal preparations remains open. It is interesting to note that for the largest peripheral rods with $d \approx 1.8\mu$ and $n_1 = 1.41$ (Polyak, 1941; O'Brien, 1951; Sidman, 1957) a value of $V = 5.5$ occurs for the shorter visible wavelengths (say $\lambda = 0.45\mu$) when $n_2 = 1.335$, a value very close to that of Enoch's immersion medium.

The importance of Enoch's observations is, in any case, undoubted. While the apparent waveguide size of the receptors he observed may be larger than can be expected *in vivo* and certainly larger than is the case for the foveal cones, his results are a convincing demonstration of low-order waveguide transmission in retinal receptors. Moreover, Enoch (1961a,b) also observed spatial separation effects along the length of the receptors whereby the mode transmitted varied as the depth of focus of the viewing optics was varied, an effect analogous to the spatial separation by wavelength in small cones when only the $HE_{11}$ mode can propagate.

# Photoabsorption Localization and Color Discrimination

In the proposed cone spectrometer model, color information is distributed along the length of the cone. Thus, a critical issue for any scheme one might employ to read out the available information is that the site of any given photoabsorption event must be localized to a finite position along the cone. If a photoabsorption event at a given position along the length of the cone were to result in a signal that reaches and closes the ion channels along the cone outer segment in a totally undifferentiated way (anywhere or everywhere along the cone length) then there would be no way to make use of the spectral information encoded along the cone length. That is, if the cone outer segment simply acts like an undifferentiated bag containing photopigment, spatially encoded color information could not be read.

The excitation signals resulting from photoabsorption at a particular location along the receptor outer segment will spread away from the absorption site due to longitudinal diffusion of cGMP, the cytoplasmic second messenger in the photoreceptor detection process. The photoreceptor outer segments have extensive internal structure (disks in rods and membrane evaginations or infoldings in cones) that can be expected to profoundly influence this longitudinal diffusion in a complex way. The details of this anatomical structure can be expected to limit this spread of cGMP significantly and in different ways in rods or cones.

Until recently there had been no direct measurements or theoretical investigations of longitudinal diffusion in cones, but there have been a number of such measurements in rods, including that of Hagins, Penn, and Yoshikami (1970), Hemila and Reyter (1981), Lamb, McNaughton, and Yau (1981), Matthews (1986) and Gray-Keller, Denk, Shraiman, and Detwiler (1999). Hagins, et al. (1970) for example found that the longitudinal diffusion of the transduction signal in the rods was confined to a length of 10 μm or less.

An elegant study by Holcman and Korenbrot (2004) directly measured longitudinal diffusion in both cones and rods and also

73

theoretically modeled the phenomena in terms of the detailed physical structure of the receptor outer segments. Their results are best summarize in a quote from their abstract:

*At a given time interval, cGMP spreads further in rod than in cone outer segments of the same dimensions. Across all species, the spatial spread of cGMP at the peak of the dim light photocurrent is 3-5 μm in rod outer segments, regardless of their absolute size. Similarly the cGMP signal spread is 0.7-1 μm in cone outer segments, independently of their dimensions.*

Thus, the site of a given photoabsorption event in a cone is localizable to within about 1 μm along its length. For 40 μm long foveal cone outer segments, this provides a potential resolution, from a single cone, of 1 part in 40 or 2.5% of its length and thus the spectral range encoded over that length. For a visual range of say, 660 to 440 nm, this 200 nm spread in wavelengths could be potentially read from a single cone with a resolution of 5 nm.

Now, human visual color discrimination is somewhat better than this as we saw in Figure 8 from the classic data of Wright and Pitt (1934). There, best wavelength discrimination is about one nm, some five times better than the upper bound inferred above. One would certainly expect that as more cones participate in detection, discrimination would improve. Assuming a square root dependence in improvement with the number of cones participating (as in signal to noise ratio improvement) then five times better discrimination would be possible with twenty-five cones participating. This number of cones would occupy twenty-five times the area of a single cone but only five times the linear extent of a single cone. With the angle subtended at the retina of a single cone on the order of about one minute of arc, the cone spectrometer mechanism would thus offer a minimum discriminable wavelength of about one nm with an image size on the retina of only about 5.0 minutes of arc. These dimensions are considerably smaller than the minimum stimulus size used in typical color discrimination measurements and thus easily realizable for human vision.

Of course, the above simplistic analysis is not the entire story on the capability of the cones to provide color discrimination. The resolution of the mechanism that might be used to read the detections encoded along the length of the cone also comes into play. The above arguments have said nothing about this read-out mechanism. We will return to this subject in terms of the temporal characteristics of color detection since we will argue that a natural time code is used to read out the color information. However, regardless of the actual mechanism used, the above considerations indicate some of the fundamental limitations of the process.

Note, by the way, that in comparing this process to the information provided in the standard three-cone model, one would only need a read-out capability of one-third the length of the cone (about 13 μm for foveal cones). That is, partitioning color information into three channels at the very first stage of detection (as in the three-cone model) requires substantially less resolution than is apparently available in the cones.

# Chapter 5
# Cone Spectrometer and Retinal Cone Function

If one set out to build a physical instrument for the resolution of color employing the color-dispersive properties of a small dielectric cone, one would have to meet specific design criteria in its construction. The color information potentially available from the device would be of a particular kind and would be limited in particular ways, regardless of how information is actually detected. In this section we examine how human retinal cone structure compares with ideal cone spectrometer design and to what extent human color vision shares the same limitations in performance characteristics as the cone spectrometer.

## Information Characteristics and Operating Range

The optical transmission characteristics of a dielectric cone only define how light can illuminate its interior and cut off curves such as Figure 15 do not say how the photon energy is actually deposited in the cone. For this one needs information on the concentration, spectral absorbance, and distribution of any photosensitive pigment within the cone. As was evident from the discussion on the single cone MSP results, these parameters are rather incompletely known. While it is thus rather difficult to define the actual pattern of energy deposition along the retinal cones there are nevertheless some general statements that can be made about the kind of color information potentially available from the cones, regardless of what kind of photopigments might be present, how they might be distributed, or the available information extracted.

In the eye, light normally enters the cone through the inner segment before being guided along the axis of the photosensitive outer segment in the direction of decreasing diameter. The incident light passes first through the broad proximal end of the segment and thus any one photon detection event there represents rather little color information, since it could be due to a photon of any wavelength. However, the information content of an absorption event increases when it occurs in increasingly more distal portions of the cone. At the distal tip of the cone only the shortest wavelength photons remain to illuminate the cone interior to a significant degree.

Thus, any light-dependent signals generated there are uniquely correlated with the shortest wavelengths for which the cone is "tuned". If it is "tuned" to the visible spectrum, then one could determine the color of blue light on the basis of very few absorption events because of this uniqueness or high information content of single absorption events in the distal portion of the cone. As a consequence, blue light would represent a relatively high ratio of chromaticity information to intensity information as compared to other colors in cone spectrometer operation.

It is interesting that a similar feature is observed in human color vision. It is well-known that blue light is, for one reason or another, detected in a somewhat different way than other colors in the retina, that it does have a higher Weber fraction (chromaticity to brightness ratio) and contributes disproportionately more to chromaticity information than do other colors (Stiles, 1959). In the multiple receptor framework this feature is usually attributed to either the relative paucity of blue cone types (Wald, 1967) or else that the blue cones are distributed normally but their outputs are grouped differently, with many blue cones contributing to a single output channel (Brindley, 1960). Actually, it is notable that a common theme throughout the current literature is that the so-called blue (or S) cones are very rare and that any given area contains very few (or even no!) blue cones (c.f., Roorda and Williams, 1999, Curcio, et al., 1990). This conclusion has come about through microspectrophotometric techniques looking for blue pigment absorption as well as differential staining techniques intended to identify these blue cones.

To the extent that color vision is not based on single absorption events the non-unique character of absorption events in the broad end of the cone does not necessarily mean that red information would be lost. As long as one has information on the pattern of energy deposition along the cone, then red light and blue light can be distinguished from each other and both can be distinguished from "white" since the illumination patterns are all different. For example, red light will illuminate the small end of the cone much more weakly than will blue or white light (containing both red and blue). Similarly, white light (which, of course, contains a red component) can illuminate the broad end of the cone proportionately more than can blue light.

On the other hand, a monochromatic light in the middle of that operating range of wavelengths for which the diameter variation in a particular cone is "tuned" will have a pattern with the least difference from that of white light. If the cone is constructed to tune over essentially the full visible spectrum (see below), then the least saturated (least different from white) wavelengths should be those in the middle range of the spectrum, somewhere around 540 nm or so, near the peak in photopic sensitivity.

These qualitative characteristics of color information more or less parallel the saturation characteristics of human color vision; the colorimetric purity curve (steps from white) for different wavelengths is basically a V-shaped function, far from white at either end of the spectrum with the least saturated color a yellow around 580 nm wavelength (Wright and Pitt, 1935).

That the yellow wavelengths are the least saturated does not mean that there is loss of resolution, however. Highest resolution in human color discrimination (minimum wavelength differences seen as different) occurs for the same yellow portion of the spectrum. The cone spectrometer's ability to sort colors is characterized by the modal cutoff curves (Figure 15). Intuitively, the greatest difference in transmission for different wavelengths must occur at the steepest portions of the cutoff curves where the efficiency changes fastest as a function of waveguide size V.

The optimum operating range for the cone spectrometer is apparent when the derivative of the cutoff curves is plotted. Figure 22 is a plot of $d\eta/dV$ as a function of V. Where the rate of change, $d\eta/dV$, is largest determines which values of V correspond to the greatest potential wavelength resolution. The very large values of $d\eta/dV$ near cutoff for the $HE_{21}$ mode (rising asymptotically to infinity) simply reflect the abruptness of cutoff at its critical value of the waveguide parameter (V = 2.405). The form of $d\eta/dV$ is different for the dominant $HE_{11}$ mode, where we expect the cones to operate because there is an inflection point in the cutoff curve at around V = 1.08 after which the efficiency decreases more slowly with V, cutting off exactly only at V = 0.

The maximum rate of change in the $HE_{11}$ modal efficiency occurs at the location of this inflection point, at which the efficiency of the mode (from the cut off curve) is approximately 23%. This

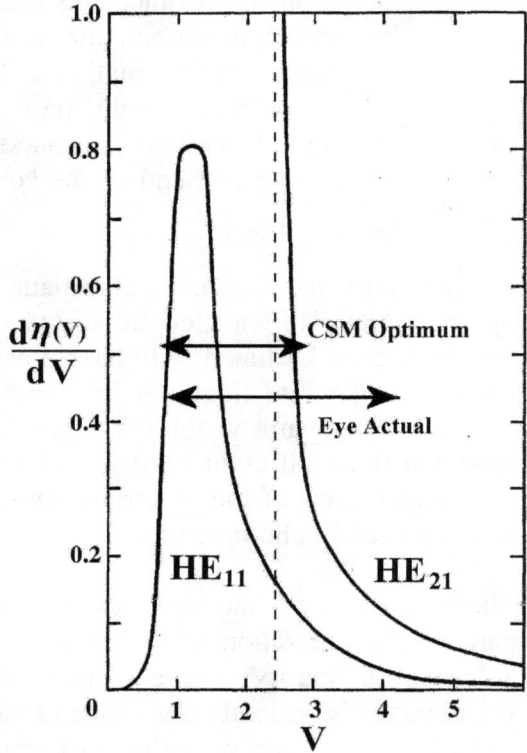

Figure 22. Operating range for cone spectrometer is indicated by plot of $d\eta/dV$ curves for the $HE_{11}$ and $HE_{21}$ modes. Large values for the rate of change in the modal efficiency correspond to potentially high resolution of wavelength differences (large change in transmission for a small change in wavelength). The range of V values (approximately 0.8 to 4.2) for the visible spectrum propagating in the retinal cones is indicated for comparison to the "ideal" range.

is perhaps better than one might have anticipated: the modal efficiency is still reasonably high where the cutoff is most rapid and potential resolution of colors is the best. What this maximum in $d\eta/dV$ means in terms of the minimum difference in wavelengths which can be resolved as different in a particular cone spectrometer will depend on the construction details of that instrument: it will depend on the physical distance over which the change in waveguide parameter is spread and the resolution with which the illumination of different portions of the cone can be read.

For a cone spectrometer, one would naturally assign the midpoint of the spectrum of wavelengths to be resolved to the maximum of the discrimination curve at $V = 1.08$. Logically one might specify this value of V for the mid-range over which the cone could operate with its range being limited on either side of this maximum. On the small V side, it is limited by both the decrease in resolution ($d\eta/dV$) and in efficiency ($\eta$) as V gets small. We may take as a reasonable limit, V $= 0.8$. For the $HE_{11}$ mode, $d\eta/dV$ also decreases on the large-V side, although when V is as large as 2.405, the next higher mode can be excited. Since the waveguide will tend to propagate in the highest order mode it is capable of supporting (Kapany and Burke, 1972, p. 231), this results in closure of the cone spectrometer operating range.

As one goes to wavelengths small enough that the waveguide parameter is 2.405 or larger in the broadest end of the cone, the transmission pattern will suddenly change. Instead of attenuating relatively slowly, as would ordinarily occur for blue light (in the $HE_{11}$ mode), upon excitation of the second order mode by sufficiently short wavelengths, the attenuation pattern will be more like that for very long wavelengths. Actually only part of the total power of the radiation field will be in the higher order mode, and within a short distance this would be cut off so that the remaining power propagates as short wavelength light in the $HE_{11}$ mode. That is, violet will illuminate the cone interior somewhat like a mixture of red plus blue light (purple). The cutoff for the $HE_{21}$ mode is very abrupt and $d\eta/dV$ is very large. Thus, the red-like portion of the violet pattern will turn on rather suddenly as one decreases the incident wavelength through blue to violet. This abruptness means that the wavelength discrimination possible at this transitional wavelength will improve abruptly as well.

The language in which this closure property of the cone spectrometer has been described bears a rather suggestive similarity to some notable features of human color perception, namely: the distinct impression of a reddish quality to violet light seen for wavelengths shorter than about 440 nm (which begins rather abruptly at that wavelength), the similarity of the sensations of violet and purple (red plus blue mixture) and the anomalous, nearly discontinuous improvement in wavelength discrimination observed at 440 nm.

For the excitation of the second order mode of the cone spectrometer to be the explanation of these features, the cones would have to meet some rather specific requirements. Explicitly, a wavelength of 440 nm in the widest portion of the retinal cones must correspond to a value of waveguide parameter of 2.405. Other requirements of the retinal cone construction implied previously are that of a long wavelength limit to the tuning range of the spectrometer due to $HE_{11}$ cutoff for $V = 0.8$ and that of the correspondence of a midrange wavelength at which the spectral discrimination is maximum with $V = 1.2$.

How well do the human retinal cones simultaneously meet these requirements? The answer is quite well, particularly in view of the simplicity of the assumptions built into the model. On this point, one must be aware of possible complications not allowed for in the absorptionless cone model. We will compare the retinal cones with ideal spectrometer operation as if the relative refractive index were independent of wavelength. This is an over-simplification, of course. All materials exhibit some refractive index dispersion and the dispersion is so-called *anomalous* for wavelengths near its optical absorption maximum. While the effect of this anomalous dispersion might (Snyder and Richmond, 1972) or might not (Stavenga and van Barneveld, 1975) significantly affect the spectral sensitivity of a receptor, there is no doubt that a small change in relative refractive index of an optical waveguide could significantly modify its transmission characteristics, particularly near mode cutoff. For example, a difference in relative refractive index as small as 0.8% could, in an extreme instance, decrease the modal efficiency by as much as 33% from $\eta = .45$ to $\eta = .30$. Since experimental knowledge of receptor refractive index is so incomplete, no corrections for index dispersion have been applied.

Even with these qualifications, however, there is so much overlap between that range of waveguide parameters required for ideal cone spectrometer operation and that defined by the best available evidence on the human retinal cones that it is consistent with the receptors having the optimum values of those parameters.

The range of V values for cones of a given size depends on the range of $\lambda$. The wavelength range to which the eye is sensitive extends at the very extreme from about 400 to 700 nm although above about 660 nm wavelength discrimination fails asymptotically and longer wavelengths are not distinguishable as different (Wright and Pitt, 1934; Bouman and Walraven, 1972). In the cone spectrometer model for color vision, 660 nm thus corresponds to the long wavelength, small-V discrimination limit at V=0.8. The short wavelength "limit" for the spectrometer (defined by closure of the operating range at V=2.405) we provisionally specify as $\lambda = 440$ nm. With these wavelength limits, the range of waveguide parameter V = $(\pi d/\lambda)$ NA for human retinal cones spans the largest and smallest values obtained respectively with $\lambda = 440$ nm light propagating in the largest diameter region of the outer segment and with $\lambda = 660$ nm light in the smallest diameter regions of the cone. For the numerical aperture we have used the (probably conservative) value of NA = 0.33.

This range of V values for human cones is indicated by the span marked in Figure 23 by the triangles for cones at different retinal eccentricities. A pair of best-fit dotted lines is used to indicate the receptor V-value range. The "ideal" range for a cone spectrometer is indicated by the region between the two dashed lines (bracketing V-values between 2.5 and 0.8). As is evident, there is a considerable amount of overlap in ideal and actual cone range.

For a more specific comparison of the ideal cone spectrometer and a representative foveal cone, consider one with an outer segment 0.9 μ at its widest, tapering to 0.6 μ. For this cone to exhibit closure of its "color circle" at 440 nm requires that it have NA = 0.38. For it to exhibit maximum discrimination at 580 nm (V = 1.2 in middle of cone) requires NA = 0.30. For its long-wavelength limit to be 660 nm requires NA = 0.28. These values compare favorably with the numerical aperture value for retinal cones of 0.33 discussed in the previous section. While the increasing

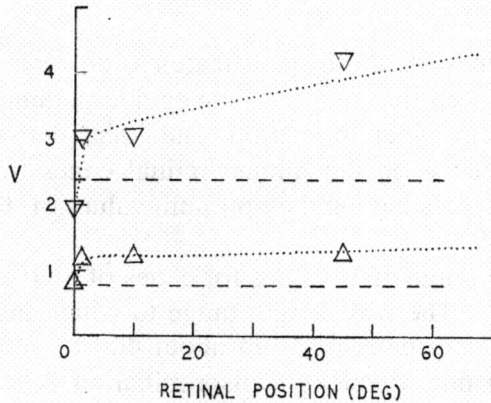

**Figure 23. Range of waveguide parameter values for light of wavelengths between 440 nm 660 nm propagating in the cones across the human retina (using the conservative estimate of NA = 0.33) bracketed by the dotted lines. The ideal range for cone spectrometer operation is indicated as bracketed by the dashed lines.**

value of numerical aperture for shorter wavelengths is suggestive of a dispersive refractive index (NA can be made equal for all three criteria with less than 2% change in relative refractive index) the exact wavelength assignments for the criteria are too approximate to precisely define that dispersion.

Such a comparison does indicate that the retinal cones are very close to the optimum for spectroscopic operation. The assignment that can be made with the greatest confidence is that of second-order mode excitation with the violet perception (color circle closure). In addition to the prediction of the qualitative features of the violet

84

sensation, the proposed model accounts for the very abrupt improvement in the human color discrimination curve at around 440 nm.

While the improvement in normal color discrimination (Wright and Pitt, 1934) at 440 nm is not of large magnitude, it is relatively abrupt. Even more significant, however, is the form of the human color discrimination curve for the abnormal conditions of strict foveal fixation (Willmer and Wright, 1945) and for tritanopic color defective vision (Wright, 1952). Color discrimination in these two conditions is very similar. This similarity has been noted previously (see Wright, 1971 for summary), usually in connection with the so-called blue-blindness of the fovea.

As we shall see subsequently, this similarity is more than coincidental and is related to the temporal coding of color information through eye movements in the proposed cone spectrometer model. For now, the significant point is that under both of these conditions there is a virtually discontinuous improvement in color discrimination. This break in the color discrimination curve occurs for both conditions at around 440 nm.

## Similarity of Violet and Purple

As pointed out in the previous section, the detection of color information in a cone spectrometer depends on resolving the location along the cone of a light absorption event. Regardless of what scheme might be employed to detect the location of absorption events, the device resolution improves with the length of cone over which the color dispersion is spread. As pointed out above, the retinal cones are longest where the color vision they provide has the greatest resolution, in the fovea. While this is not the entire story -- the number density of cones decreases and synaptic linkages become more rudimentary with increasingly peripheral retinal position -- the marked correlation of cone length and color discrimination over the extent of the retina is not easily explained otherwise. However, if the cones utilize some of the information available in the spectral dispersion that must occur along them, then the improvement in resolution with increased cone length is simply explained.

A second apparent parallel between retinal cones and cone spectrometers can be found in the way they both tend to confuse violet and purple. When one observes the output of a tunable monochromatic light source, as its output wavelength is decreased below about 500 nm, the blue appearance becomes violet at a characteristic wavelength. The exact wavelength for the onset of the violet appearance depends on the observer and the intensity of the light. This violet point usually ranges from 480 to 430 nm: a typical value is about 450 nm. The transformation of blue into violet is accompanied by the distinct impression that red light is being added to the blue. This is, of course, why it is so easily confused with purple, which is an actual mixture of red and blue.

Previous attempts to explain this phenomenon have assumed the existence of a subsidiary maximum in the absorption spectrum of the photopigment presumed to be present in the "red cones." Thus, because of the relatively greater absorption of light of short enough wavelength, these red cones would be excited in addition to the blue cones, thereby explaining the similar response to violet and purple. However, no such short-wavelength sub maximum has ever been resolved in any of the MSP measurements on single primate cones nor in the L-pigment spectral absorption curves from the molecular genetics studies. Indeed, it has proved difficult to find any blue-absorbing photopigment at all in the primate retina. It is also notable that the absorption peak of the photopigment detected in the "red cones" is not in the red at all, but in the yellow-green at about 565 nm. There thus appears to be no satisfactory explanation in terms of observed photopigments alone, for the similarity of violet and purple and the consequent closure of the color circle in human vision.

If retinal cones utilize the spectral dispersion produced by mode cutoff over their length (in combination with any single - or multiple - photopigment that might be present), then the red appearance taken on by short wavelength light would be an expected consequence of second-order mode propagation. For decreasing incident wavelength, the waveguide mode parameter will increase: when it exceeds 2.405, second-order mode propagation will be allowed in the entrance end of the cones. As discussed in the previous section, these second order modes will cut off rapidly and mimic the pattern produced by red light for the $HE_{11}$ mode. Inasmuch as only a fraction (depending on the input coupling conditions) of the incident short wavelength light will propagate in second-order modes, while the remainder propagates in the first-order $HE_{11}$ mode, this pattern would be difficult to distinguish from one produced by a blue and red light mixture propagating in the $HE_{11}$ mode alone.

Given this explanation of the similar appearance of violet and purple, one should be able to predict the wavelength of the violet point in human vision in terms of the physical parameters of the retinal cones. As discussed above, the foveal cone diameter at the entrance end of the outer segment is about 1.0 μm. Using the best available estimates and measurements of refractive index, the retinal cone waveguide parameter is approximately given by $V = d/\lambda$. These values predict a violet point ($V = 2.405$) for an incident wavelength of 416 nm. For the violet point to occur at the typically observed value of 440 nm, V must be $1.058 \ d/\lambda$. Using $n_1 = 1.387$, this requires that $n_2 = 1.3454$. This value for the interstitial medium is less than the commonly "accepted" estimate of $n_2 = 1.347$ by less than 0.2%.

Thus a brief review of the information available on the physical parameters of the primate retinal cones suggests that the visible spectrum is dispersed along their length by low-order mode cutoff. An examination of some basic properties of color perception suggests that this mechanism may play a functional role in human color vision. The clear implication of these results is that much more attention should be focused on the physical parameters and anatomy of the human retina than has hitherto been the case. A closer examination of the relation between cone structure and cone function may well lead to increased understanding of human color vision. It may also suggest design concepts and ideas for the construction of cone spectrometers of practical utility.

## Directional Properties – the SC Color Change

Modern day interest in the optical guiding characteristics of retinal receptors arose in connection with the discovery of the directional sensitivity of the retina (Stiles and Crawford, 1933). That the apparent brightness of a stimulus decreases with increasing angular deviation from nominal incidence, the Stiles Crawford effect of the first kind (or SC I), is a manifestation in some form of optical guiding in the receptors is no longer seriously doubted.

There is an analogous directional effect in color perception, the Stiles-Crawford effect of the second kind (SC II), first demonstrated by Stiles (1937). In this effect, the perceived color of a stimulus alters as its angle of incidence at the retina is altered, primarily shifting to longer wavelengths with increasing angle. Explanations of this color shift in terms of pigment self-screening effects in a three-photopigment model have been proposed (Walraven and Bouman, 1960; Wijngaard, Bouman, and Budding, 1974). This explanation postulates directional color shifts arising through the preferentially greater absorption with greater path lengths through a pigment, which occurs for light of those wavelengths near the pigment absorption maxima. While with a sufficiently large number of free parameters to choose from (wavelengths of three photopigment absorption maxima, photopigment densities and distribution), one can mimic the observed directional color shift with this approach; the pigment densities one must postulate are improbably large (Enoch and Stiles, 1961). In a review of the SCII effect, Alpern (1986) concluded "that 'self-screening' theory may not provide a satisfactory description of the color changes throughout the visible spectrum".

An effect similar to the SC II color shift for transmitted light was observed in bleached retinal preparations by Enoch (1961b) where "The changes in modal pattern induced by oblique irradiation result in many instances in the same physical distributions of energy which are obtained by increasing wavelength." That this effect was observed in bleached preparations requires an explanation other than pigment self-screening in this case.

To our knowledge it has not previously been noted in connection with the directional color shift that a basic wavelength-dependent directional property of an optical waveguide is just the red shift observed as the basic feature of the physiological effect (and in Enoch's retinal preparation). We previously proposed an explanation of the SC II effect on just this basis (Medeiros, 1979). In a dielectric fiber, the propagation constant determining how light is guided in the fiber specifically depends not on the wave vector $\mathbf{k}$ itself, but rather its component in the z-direction of the guide axis given by $\mathbf{k} \bullet \mathbf{z} = k\cos\theta$ (c.f. Kapany and Burke, 1972). Thus light of physical wavelength $\lambda$ (with wave number $k_1 = (2\pi n_1/\lambda)$) incident at an angle $\theta$ to the guide axis can be thought of as equivalent to light propagating axially but with an effective or guide wavelength given by $\lambda_g = (\lambda/\cos\theta)$. The predicted shift in $\lambda_g$ is thus always to longer wavelengths, i.e., a red shift.

The simple $\lambda/\cos\theta$ function is directly compared to the SC II data of Stiles (1937) in Figure 24. For most wavelengths this simple function is a remarkably good fit. Even where it does depart from this form (blue-shifting at large angles for $\lambda$ between 570 and 500 nm) it follows the $\lambda/\cos\theta$ form at small angles in all cases.

Now, the $\lambda/\cos\theta$ form for guide wavelength is a general property of optical waveguides and thus any color vision model based on some wavelength-dependent feature of optical transmission (cone classes of different diameter, detection of different waveguide mode patterns, or the present model) would predict the same basic form for the color effect. Where different models must differ is in their explanation of the deviations from the basic $\lambda/\cos\theta$ form in the middle wavelength range for the larger angles.

89

The $\lambda/\cos\theta$ dependence is the color shift for a single isolated optical fiber. The human retina is, however, a closely packed assembly of receptors. Since light incident on a fiber at sufficiently extreme angles can escape to the external medium, then in the retina this escaping radiation can couple into surrounding receptors. The escaping light is downward directed and is incident on the neighboring cones at a more distal region of it than that of the receptor from which it escaped. That there is thus more light incident more distally on the receptors than is normally the case for axially incident light would result in a blue shift for large angles since an increase in the signal generated in the narrow portions of the cone is associated with short wavelengths.

This blue-shift effect at large angles would be

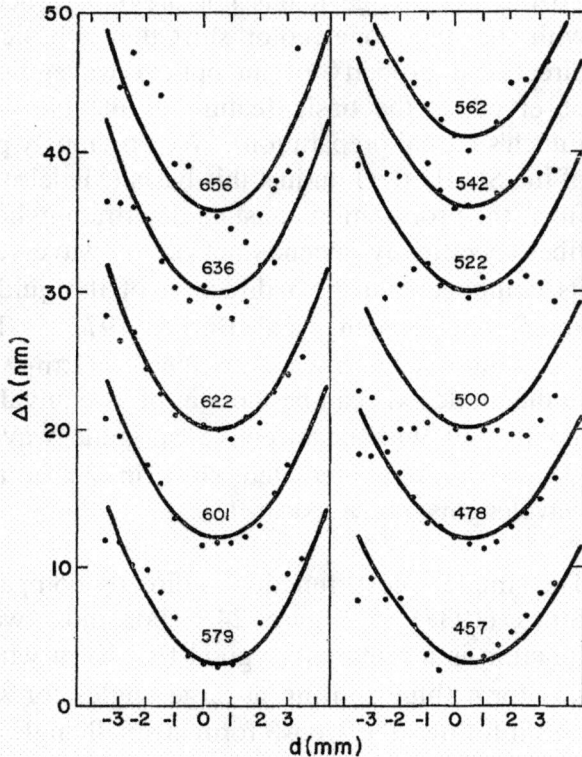

Figure 24. The Stiles-Crawford color change for light incident from different directions at the human retina (light entering at an eccentric position d mm from the pupil center). The original data of Stiles, 1937 (black dots) is compared to the predicted color shift for a single isolated cone spectrometer (black line). Results for different test wavelengths are arbitrarily displaced vertically for clarity. The fit of the simple guide wavelength function to the data is quite good and only deviates significantly at the larger angles for the test wavelengths between 542 and 500 nm.

manifest only for the proposed cone spectrometer model. The details of this light leakage at large angles are rather difficult to compute because of the geometrical complexity of the total path length through the receptors (additionally involving the details of the inner segment). Whatever its details, however, the coupling to neighboring cones will be greatest for light of those wavelengths near the absorption maximum of the photopigment contained in that part of the cone on which the sideways-directed light is incident. This is obviously fertile ground for further research on more detailed measurements of the SC II color change effect and more detailed theoretical calculations taking into account the detailed geometry of the photoreceptor matrix, coupling into adjacent cones, and photopigment absorption characteristics.

## Chromatic Adaptation

Another illustration of the potential usefulness of the proposed model concept is the ready facility with which it can account for some heretofore rather puzzling observations on the adaptational properties of human color vision. In an important series of experiments Brindley (1953) explored some of the effects on human color vision of adaptation to very bright spectral lights. He found that adaptation to a bright yellow light of 578 nm dominant wavelength disturbed the amount of red necessary to match lights of different wavelengths much more than it did the amount of green necessary for that match.

Moreover, this difference in the way the red and green components necessary for a match were disturbed by adaptation is essentially independent of the wavelength of the adapting light. As Brindley noted: "If 2 or more different processes were concerned in the disturbance of color matches by adaptation, e.g. if the form of the spectral sensitivity curves of two or more classes of receptors were altered, the ratio $\Delta LogR/\Delta LogG$ would not necessarily be the same for different adapting wavelengths; indeed we should expect it to be positive for some and negative for others. But if the disturbance is due to a single process with a unique spectral sensitivity curve, then the constancy of ΔLogR/ΔLogG is explained." The proposed cone spectrometer model provides just such a single process.

Brindley suggested that the concept of multiple receptors could be retained and his results explained if the "red cones" were predominantly affected by the adaptation. Thus he proposed that the density of red pigment was some five times greater than that of the green pigment. It should be noted that no such huge difference in density of red and green pigments has been indicated in the indirect bleaching measurements of reflection densitometry or microspectrophotometry.

Adaptation to a combination of violet ($\lambda = 438$ nm) and red ($\lambda = 658$) light produced a state in which any light with a wavelength between 480 and 625 nm could be matched by 578 nm light alone. The sensation mediated was that of an *unsaturated blue-green.* Adaptation to yellow with $\lambda = 578$ nm produced a very distinctive state in which light of any wavelength less than 500 nm could be matched by a light of 447 nm wavelength. The sensation mediated was that of a *very saturated violet,* substantially more saturated than any spectral violet seen with the unadapted eye.

This result is sensibly what is expected in the proposed model. In the "isolation" of the first two mechanisms (especially in the case of the "red mechanism") the adaptation involves exposure to light of wavelengths shorter than those with which that portion of the cone is correlated. Light is illuminating the cone along its entire length, since short wavelengths are included in the adapting light, and thus adapting or "fatiguing" the cone over that full length including the region "isolated".

In the case of isolation of the blue mechanism, however, the situation is qualitatively different since little adapting light has reached the distal tip of the cone to "fatigue" it. Thus, the portion of the cone correlated with the detection of short wavelengths is relatively unaffected and thus can contribute to color sensation relatively more strongly than is the case of the other "isolated" mechanisms. That is, the sensation in the third case will be "purer" or more saturated; indeed we should expect it to be more saturated than in normal conditions as well since only the isolated distal tip of the cone is responding in contrast to normal vision where the rest of the cone is not adapted and signals are produced along its full length.

## Colorimetric Purity and the Saturation of Colors

The issue of color saturation and colorimetric purity was discussed in Chapter 1 as one way of describing the trichromacy of vision in terms of the hue, saturation, and brightness. The data of Kraft and Werner (1999) was plotted in Figure 9 displaying the typical V-shaped pattern as a function of wavelength. We might intuitively expect that this sort of functional dependence on wavelength to make sense in the cone spectrometer model since yellow, in the middle of the spectral range would illuminate the cone interior most like white light. Red, attenuating quickly would seem to be very different from white, and blue, uniquely reaching the distal end of the cone, should also be very different from white.

To properly compute colorimetric purity in the cone spectrometer model, we would need data on the exact dimensions and refractive indices of the cone and its surround to model waveguide mode attenuation along the cone. Even more, to determine how different spectral lights are absorbed, one would need data for any resident photopigments and their spatial distribution within the cones, their optical density, and their spectral absorption curves. We would further need a good model of how the spatially-ordered information is read out and coded in the visual process. In lieu of having the needed details to accurately compute this, we can make some very crude simplifying approximations and attempt to see if the cone spectrometer model could even plausibly lead to the kind of colorimetric purity function actually observed.

To do this we use a cone model with diameters ranging from 1.4 to 0.4 um and a numerical aperture of 0.9 and assume here that all colors are absorbed equally with no assumptions about how any

93

photopigments might be distributed within the cones. We then use a white made up of equal mixtures of red (690 nm), yellow (580 nm) and blue (440 nm). We further only compute the spectral distribution along the cone in this simple model for only the $HE_{11}$ mode. For this very simple and crude approximation, we plot in Figure 25 the difference from white of three spectral colors (690, 580, and 440 nm) as the difference in distribution of each color from that of a white normalized to have the same intensity.

The difference curves for red (upper curve), yellow (middle curve), and blue (lower curve) wavelengths are plotted in the figure. Note that, just as one might intuitively expect, yellow indeed has the least difference from white (note that zero difference from white is in the middle of the graph), and the red (positive difference) and blue (negative difference) are more different than white.

**Cone Spectrometer Model**
**Difference from White**

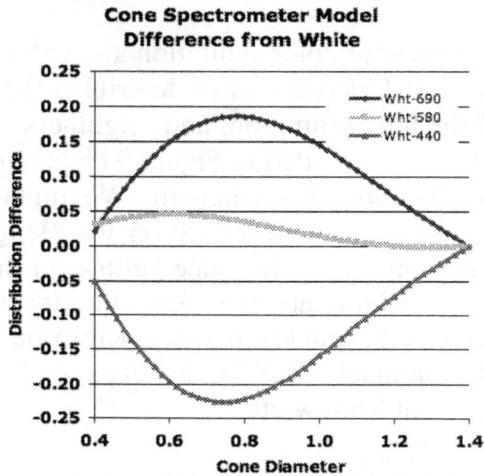

Legend: Wht-690, Wht-580, Wht-440

**Figure 25. Cone spectrometer difference from white for 3 colors – upper curve for red, middle yellow, and lower blue**

It is interesting that, even in this very simple model, the blue is actually even more different from white than is the red.

A simple integration of the area under these curves would give a total measure of difference from white for each wavelength. Figure 26 is a plot of these integrated values as a function of wavelength where we have also included the results for additional colors at 640, 540, and 500 nm. Note that this plot for the cone spectrometer model, even under these very crude simplifying assumptions results in a V-shaped function much like the experimental data on the saturation of colors for human vision as was plotted in Figure 9 in Chapter 1.

Again, the assumptions we have used for these computations are too crude to accurately represent what would be the case for a real observer since accurate cone structure, pigment absorption and pigment distribution data are not included. However, it does show the impressive predictive power of the cone spectrometer model. Hopefully, we can look forward to even more accurate computations once better data is at hand.

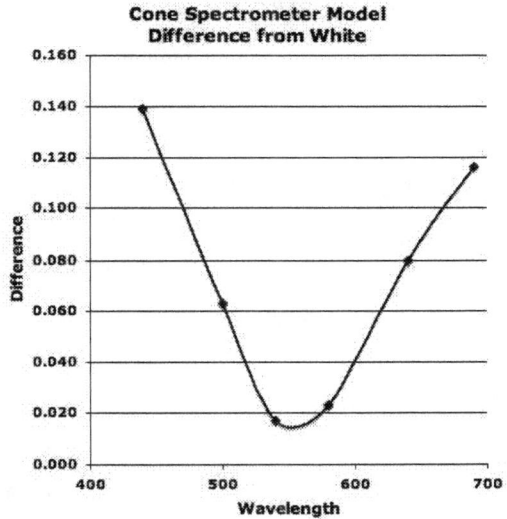

**Figure 26. Cone spectrometer spectral purity modeling**

As in the assumptions we have used for these computations are too
to estimate based on this model in the case for a real
material since accurate data indicate pigment absorption and
reconstruction data are not in the illustration. If these show the
impressive glue curve power of the modeler more accurate model
As a result does I expect to even more accurate complications
the reader else is in mind.

Figure 21. Comparative prey spectral
sanity assuming a ... Figure 12. From.

# Chapter 6

# Temporal Coding and Color Information in the Cones

The primary concern of this chapter is the temporal coding of color information in the retinal cones. This, we will contend is the logical way to encode and read out the color information generated by the cone spectrometer function of the retinal cones. There is a natural mechanism present in the eye to carry out this encoding. Saccadic eye movements, positioning new inputs to the retinal cones across color borders provides a natural time reference. We will present direct data that shows that there is a correlation between wavelength of the incident light and the time delay in the resulting signal. Furthermore, as we shall see, the phenomenon of subjective color – color perception induced by temporally modulated achromatic illumination – is directly explained as a consequence of this natural temporal code for color vision.

Before getting into the details of this temporal code, however, we first discuss a separate set of experiments that were conducted by my colleagues and me. This set of experiments directly and unequivocally measured the temporal dependence of color perception on the wavelength of the inducing light. This was done by using the rod photoreceptor response as a time reference. Since the time course of the rod response was demonstrably independent of wavelength, it provided a means of unequivocally measuring the time course of the cone response. This was possible because this experiment directly – and to our knowledge – for the first time in the reported literature distinguished the separated perception of the rod and cone receptors over the full range of photopic to scotopic illumination conditions.

It should be noted that we undertook these experiments in an effort to replicate the research of Ives in which he directly saw the breakdown of metameric matches. As we discussed earlier in this volume, he was able to show that a yellow produced from a mixture of red and green that matched a pure monochromatic yellow under static conditions, was distinctly different when viewed under dynamic conditions. The point of course, is that in a three-cone model of color vision, this cannot happen. That it does, clearly contradicts this widely accepted, mainstream model of human color vision. We did indeed observe the Ives version of the breakdown of static metameric matches under dynamic conditions where a mixed yellow separated into a leading red edge and a trailing green edge while a pure yellow remained yellow throughout. However, in the process of confirming the Ives' result, we actually found much more than we bargained for and that is what is described next.

## Separating Rod and Cone Perception

It is well known that the eye does not respond with perfect fidelity to a transient visual stimulus. In addition to some presumptive delay in the onset and offset of the visual response to a pulse of light, an attentive observer may also see an entire series of positive and negative afterimages. The existence of such afterimages has been known for a long time (McDougall, 1904a; Allen, 1926; Coltheart, 1980). Indeed, they may be readily encountered in everyday experience as a result of flash photography or accidental viewing of the sun or some other bright source of light. The afterimages seen in a given instance depend on a number of factors such as the adaptational state and attentiveness of the observer and the intensity, color, and retinal location of the visual stimulus. Afterimages from intense flashes of light may persist for a considerable period of time.

One of the best known and most studied of these afterimages is commonly referred to as Bidwell's Ghost (Bidwell, 1901). This afterimage is, under the right conditions, seen some 200 to 400 milliseconds after a visual stimulus is shut off. The effect has been "discovered" a number of times since the early nineteenth century and has variously been called Hamaker's Satellite (Hamaker, 1899), the Pursuant Image, and the Purkinje Afterimage (Judd, 1927; Karwoski and Warrener, 1942). The names by which the effect has gone reflect the various conditions under which it may be elicited.   For example, it can be readily observed as a separate image trailing at some distance behind a moving light source as the eye fixes on an unmoving reference point (suggesting pursuit and satellite).    The effect may also be observed as a transient reappearance of a stationary image a short time after the original stimulus is extinguished (suggesting afterimage and ghost).   There has been a recent revival of interest in Bidwell's Ghost because of a possible match in its time scale with that of iconic memory effects - an apparent afterimage-like effect that has been a topic of intense interest in more recent literature (Adelson, 1978; Kriegman and Biederman, 1980; Long, 1980; Sakitt, 1976; Sperling, 1960).

The main focus of the present investigation is not, however, Bidwell's Ghost, but another afterimage-like effect we have studied that appears 20 to 50 milliseconds after the original stimulus.   The time scale for this effect is nearly an order of magnitude less than that for Bidwell's Ghost and the effect is, in our observations, easily distinguished from it.   We mention Bidwell's Ghost here both to put the present study in context and to avoid any possible confusion as to the identity of the effect we have observed and will describe here.    The (as yet unnamed) afterimage effect we have studied was noted and to some extent explored in the late nineteenth and early twentieth centuries. However, the existence of this effect appears to be virtually unknown in current vision research (but see Medeiros, Caudle, and Schildt, 1982).

When one fixates on a dim reference point within a dark field of view while a small, vertical bar of light is moved across the horizontal visual field, the moving light will be seen to be closely followed by a faint whitish shimmer.   It is necessary to observe the moving light with a fixed, averted gaze since the shimmer disappears either when one follows the motion of the light or when it moves across the central part of the visual field.    Under these specific conditions one observes a colored bar of rather saturated hue closely followed by a pale, whitish bar with a shimmery appearance within an otherwise dark field.

That the trailing, shimmering image is white (regardless of the color of the inducing light), that it must be observed in a dark field, and that it does not appear to be present in the foveal area of the retina all suggest that the shimmer is associated with the rod photoreceptors. Indeed, even the variation in the appearance of the shimmer over different areas of the retina is consistent with a rod-like nature. When the effect is observed in more peripheral portions of the retina, the shimmer has a distinctly more coarse-grained appearance then when it is imaged within central regions of retina, nearer to the fovea. It is well known that while the proportion of rod to cone photoreceptors increases for more peripheral regions of the retina, the coarseness of the rod neuronal output network also increases.   In essence, in the peripheral retina, resolution is traded for sensitivity by integrating rod outputs over a larger retinal area into a single output. This physical organization of rods within the retina is entirely consistent with the shimmering appearance of the afterimage effect and the variation in the shimmer with retinal location.

This shimmering afterimage-like effect, what one might call the *Following Shimmer*, was noted in the literature as early as 1887 by Szily in Germany and Sanford (1903) in the United States in connection with a form of the so-called Fluttering Heart Phenomenon seen under low light level conditions.   The effect was also observed by von Kries in 1899 in Germany (c.f., von Kries, 1905) and McDougall (1904b & 1904c) in England.   Von Kreis noted the rod-like properties of the shimmer and suggested that it was due to direct perception by the rod photoreceptors.   He also noted that this effect was distinct from a longer-delayed afterimage (Bidwell's Ghost) that appeared to have mixed rod and cone properties. Although there have subsequently been several extensive investigations of the Bidwell's Ghost afterimage (Frölich, 1921 &1922; Karwoski and Crook, 1937), there has apparently been no further investigation of the following shimmer effect.

We have carried out a series of experiments designed to elicit some of the characteristics of this effect.   These experiments were carried in my laboratory in the Physics Department of the Pennsylvania State University, York in 1982. While there has been a rather long delay in bringing the full results of this study to light, our results quite clearly show that the following shimmer is indeed the direct response of the rod photoreceptors and not an afterimage as such.

We have examined and defined the area of the retina in which the effect is present, the relative spectral sensitivity of the eye for the presence of the effect, and its apparent visibility as a function of dark adaptation. In all respects the following shimmer has distinctively scotopic, rod-like properties rather than photopic cone-like characteristics. Thus, under the conditions of this experiment, we have been able to separate the primary response of the two photoreceptor types of the eye. Consequently, it is possible to directly address some long-standing questions about how the two receptor systems operate.

One such question concerns the extent of the rod involvement in the perception of the color blue. There has long been speculation concerning the mechanism of blue perception because of known peculiarities about how blue is perceived compared to the other colors of the spectrum. Some of that speculation includes the suggestion that rods are involved in mediating the perception of blue. However, our results show that the rods are not directly involved in the perception of blue since the following shimmer was always observed to be a colorless white, irrespective of the color of the inducing light. Furthermore, the cone response (when separated from the rod response) to short wavelength light is directly seen to be blue. Indeed, the direct cone response for lights of various wavelengths was seen to be that of perceived colors which were more saturated than normally seen with a fixed and unmoving light source. Apparently the whitish response of the rods desaturates the cone response under normal conditions.

Another long-standing question that can be addressed is that of whether or not both rods and cones are active at the same time. That visual perception shifts from being based on cones at high light levels to being based on rods at low light levels is the well-known course of dark adaptation (and the source of the so-called Purkinje shift in spectral sensitivity with dark adaptation). Presumably there are intermediate light levels (mesopic) where both systems are operative. However, it has been difficult to determine whether the rods operate at higher, photopic light levels or are disabled as part of the normal course of adaptation to high light levels. This issue is addressed in this study through measurement of the thresholds of visibility of the shimmer and of the leading colored bar. The changes in these two thresholds clearly resolve the two components of the classic dark adaptation curve. The rod-shimmer response is detectable (albeit with difficulty) very early in the dark adaptation process. Apparently, the rods are indeed functional at high light levels.

Perhaps the most intriguing aspect of the following shimmer effect is that it can be used as a tool to explore the characteristics of photoreceptor function. We have utilized the separated perception of the two photoreceptor systems to measure the temporal response of the cones as a function of the color of the incident light. In this experiment, we exploit the fact that the temporal response of the shimmer is demonstrably independent of the color of the inducing light. That is, the timing of the rod response to lights of different color is independent of the inducing wavelength. The shimmer can thus be used as a fixed reference to explore the time dependence of the cone system response. We have utilized the following shimmer effect in just this manner in order to explore some historically vexing questions about the chromatic latency of the cone photoreceptors.

**The Technique**
There are two classic methods by which afterimages have been studied. One is to simply have an observer stare at a temporally modulated light that is fixed in space -- a flash of light, for example. A second method is to have an observer track a light that is modulated in space but fixed in time (always on) -- a moving light, for example. Because of the relatively short time scale of the shimmer effect, this afterimage-like effect (unlike the Bidwell's Ghost effect) is undetectable for spatially fixed, temporally modulated light. It is,

however, readily evoked for a moving light if the proper conditions are provided.

In order to elicit the shimmer afterimage effect it is necessary to have some very specific conditions. These include a dark field of view with the observer dark adapted for several minutes. It is also necessary to have a dim, unmoving point of light in the field of view to serve as a fixation point for the observer. An attentive observer, using averted vision, may then detect a pale, shimmering glow trailing behind any small patch of light moving at a suitable speed across the visual field.

A bar of light moving in a direction perpendicular to its long dimension proves to be ideal for observing the following shimmer effect. Such a moving bar or slit of light will be observed to be trailed by a faint whitish shimmer under the specified conditions. The distance by which the shimmer trails behind the inducing light depends on the speed of movement of the bar. For a given speed the separation of the shimmer and the leading colored bar will depend on the color and intensity of the light.

The appearance of this effect is schematically illustrated in Figure 27. A dim red fixation light is in the center of a dark field and a pair of slits (red above blue) is shown on the right moving towards the left and on the left moving towards the right. The appearance is always of the colored slit preceding the whitish, shimmering slit image.

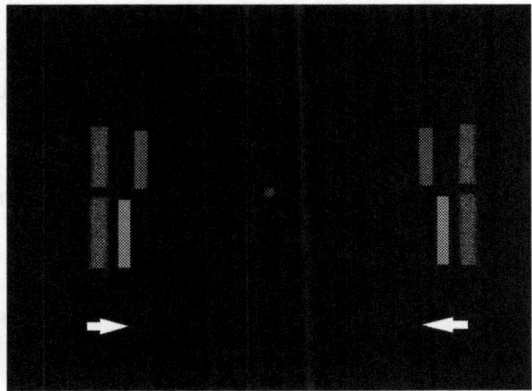

**Figure 27. Subjective appearance of the following shimmer effect for pairs of slits moving either left or right across the visual field. The upper slit is red and the lower is blue and both are trailed by their whitish shimmer. The arrows are used only to indicate the direction of motion and are not part of the observed field.**

103

The slit source positions for this image are set so that when static, the red and blue slits are directly in-line. Note here that the top, red-colored bar has the greatest precedence when moving (see data below).

The optical layout of the apparatus we used to explore the following shimmer effect is schematically illustrated in Figure 28. As illumination sources, we used the images of monochromator slits. This provided a light of the appropriate geometry (a narrow bar) whose intensity and wavelength content could be easily varied. The separate monochromator slit images are combined at a neutral beam splitter in a tip-tilt mount. The observer could vary the apparent spatial displacement of one slit image relative to the other by remote control cables attached to the tip-tilt drive motions.

The slit images are relayed along the optical system by a pair of high quality, matched transfer lenses. The images are then reflected by a mirror mounted on the rotational axis of a galvanometer movement driven by a variable electrical oscillator. After the oscillating mirror, a second beam splitter permits the superposition of a fixation point in the observer's field of view. The same beam splitter also directs the light from the monochromator slits into an imaging system which focuses the slit image onto a matching slit in front of a photomultiplier tube (PMT) detector. The PMT output, after suitable

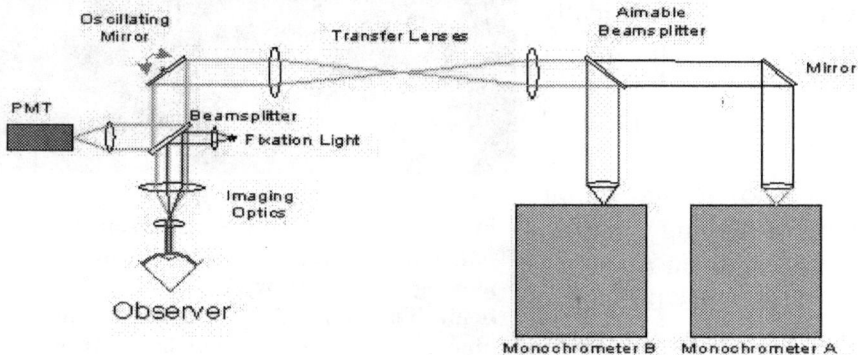

**Figure 28. Dual monochromator setup for oscillating slits.**

104

amplification, is input into an electronic timer. As the slit images are scanned back and forth, the detector outputs pulses as the slit images pass over its entrance aperture. The timer is set to measure the delay between successive pulses.

In order to measure time delays in our experiments we used two bars of light spatially displaced from each other along the length of their long axis. The observer is then allowed to physically displace one of the bars of light with respect to the other in a direction perpendicular to the long dimension of the bars (in the same direction as their common motion across the observer's field-of-view). Thus, one bar can be used as a pointer or reference while the apparent time delay of the other bar can be measured with respect to it. Imaging optics are used to present the slit images to the observer. A chin and forehead rest is provided to keep the observer's head steady to aid in proper fixation.

**Five Experiments**

Five separate experiments were carried out in order to explore the nature of the shimmer afterimage effect. The first step is to determine the nature of the effect. Is it indeed due to direct perception by the rods? The fact that the coarseness of the shimming appearance of the afterimage increases when it occurs in increasingly peripheral portions of the retina certainly is in line with this concept. However, specific quantitative data are needed to confirm the rod origin of the shimmer.

First, three experiments were conducted in order to determine if the shimmer is indeed due to direct perception by the rods. These three experiments were: (I) exploration of the retinal locus of the shimmer, (II) determination of its dark adaptation properties, and (III) verification of the wavelength independence of the temporal onset of the shimmer. Another two experiments were designed to take advantage of the results established in the first three in order to explore some of the properties of cone perception. These were: (IV) measurement of the relative chromatic latency of color perception, and (V) exploration of the intensity dependence of cone latency.


### Experiment I -- Retinal locus of the effect

It is readily noticed that the shimmer disappears when the inducing bar of light is tracked across the very central portion of the retina. One experiment concentrated on the measurement of this shimmer-free region. A second fixation light whose position in the field of view was under observer control was used as a fixed marker to indicate the boundary of the regions where the shimmer would appear to vanish. This shimmer-free region is then compared with the known rod-free or rod-poor region of the retina. A high correlation of these two regions is a necessary (although not in itself a sufficient condition) for the shimmer to be the direct perception by the rods. That the perception of shimmer occurs only for areas of the retina where rods are present is not by itself a sufficient condition for the effect to be due directly to the rods since it does not rule out the possibility of some kind of cone-rod interaction (as has been implicated for Bidwell's Ghost (c.f., DiLollo and Dixon, 1988; Sakitt and Long, 1978).

The actual experiment was rather difficult for the observer to conduct reliably. This was due to the fact that the disappearance of the shimmer towards the central retinal area was rather gradual and no abrupt, distinct boundary where the shimmer disappeared was evident. However, repeated measurements for three observers (JM, GC, and NS) placed the visual diameter of this boundary to be between $1°$ and $1.5°$ ($1.25° \pm .25°$). This corresponds well with the observed distribution of the rod-free area of the human retina, variously quoted in anatomical studies as having a diameter between $1°$ and $2.5°$ (Polyak, 1941; Yamada, 1969; Hendrickson and Youdelis, 1984; Ahnelt, Kolb, and Pflug, 1987; Curcio, et al., 1990). A value between $1°$ and $1.5°$ seems to be a consensus value defining this red-free region in the various studies, a value that agrees well with our observation of the shimmer-free region.

The approximate appearance of the shimmer as it crosses the foveal area of the observer's retina is illustrated in Figure 29. Here we show a blue-colored slit on the left moving to the right with its shimmer disappearing as it enters the foveal area. A similar effect is observed for a red-colored slit approaching from the right. We did observe that the area of shimmer-free region was independent of the color of the inducing light.

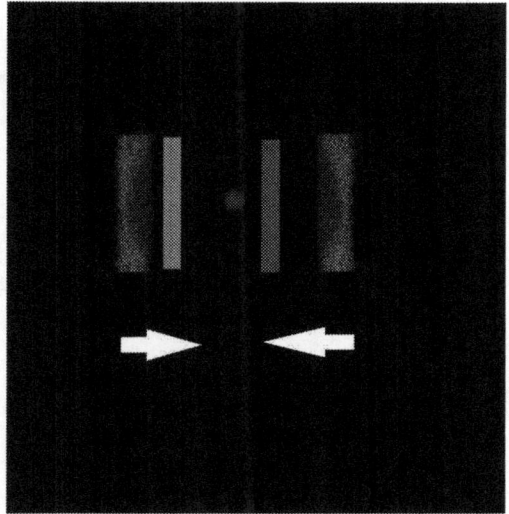

**Figure 29. Appearance of the shimmer near the fovea.**

## Experiment II -- Dark adaptation

Another means of identifying the nature of the shimmer (and perhaps differentiating rod and cone effects) is to measure the threshold of light intensity that induces the effect as the eye is allowed to dark-adapt. If the shimmer is due to rod perception, then its threshold of visibility should mimic the course of dark adaptation for low light levels. The threshold of light intensity which allows the observation of the colored portion of the bar image itself should follow what might be expected for cones under dark adaptation. That is, if one measured the separate thresholds of light intensity required for the appearance of the colored leading bar and the trailing shimmer, these threshold levels should correspond to that of the two separate receptor thresholds which (presumably) make up the normal course of dark adaptation.

The dark adaptation curves typically reported in the literature display

107

the intensity threshold for visibility as a function of time as adaptation progresses and typically display a distinct break in its curvature at around ten minutes of dark adaptation. Typical results (Bartlett, 1965) are shown in Figure 30. This break in curvature is ascribed to the response of the separate cone and rod photoreceptor systems with the cones dominating the early part of the curve and rods dominating after the ten-minute delay. Any differences in spectral sensitivity of the threshold of detection of the colored leading part of the bar image and its following shimmer should also correspond to that for cones and rods, respectively.

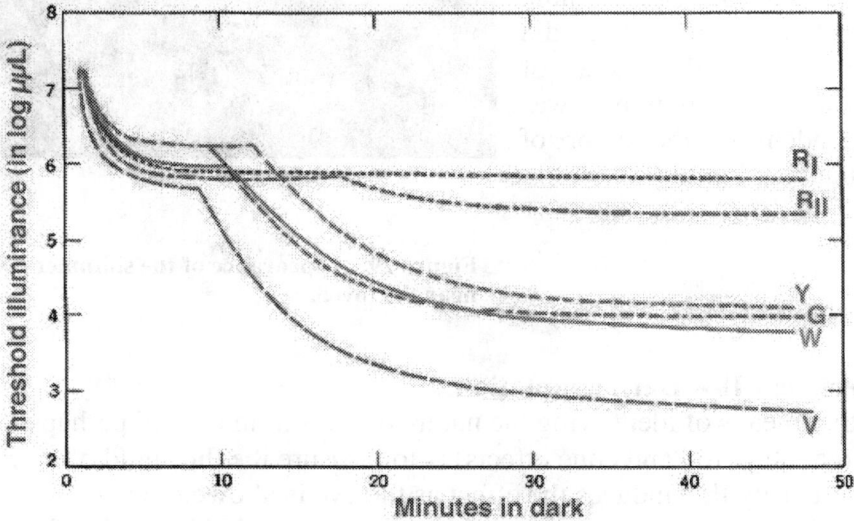

Figure 30. Classic measurements of dark adaptation in the retina showing the cone-rod "break" for different colors. Data of Chapanis in Bartlett (1965). The colors tested were $R_I$=680 nm, $R_{II}$=635 nm, Y=573 nm, G=520 nm, V=485 nm, and W=white.

In our setup, the observer was initially light adapted to bright room illumination. The experiment then began and the clock started when all illumination was extinguished. The observer decreased the slit illumination until the colored leading bar just disappeared (threshold of visibility) and the PMT intensity reading was then recorded. The observer then continued to decrease the slit intensity until the shimmer also just disappeared to the threshold of visibility where the PMT reading and the clock time was again recorded. Note that under these conditions, the observer could see the trailing shimmer without its leading colored slit. This gives the rather spooky impression of having an afterimage visible with no inducing image! Of course, with the interpretation that these are separate perceptions by the rod and cone systems, there is no apparent paradox.

This sequence of threshold determinations was continued until there was little apparent increase in sensitivity for both thresholds, typically at 1700 to 2400 seconds (28 to 40 minutes). This sequence was carried out at four test wavelengths, 450 nm, 500 nm, 550 nm, and 600 nm.

Figure 31 shows the results of this dark adaptation process for 450 nm in a semi-logarithmic plot (log of intensity versus time) for one observer (JM). The two sequences of measurements for the colored

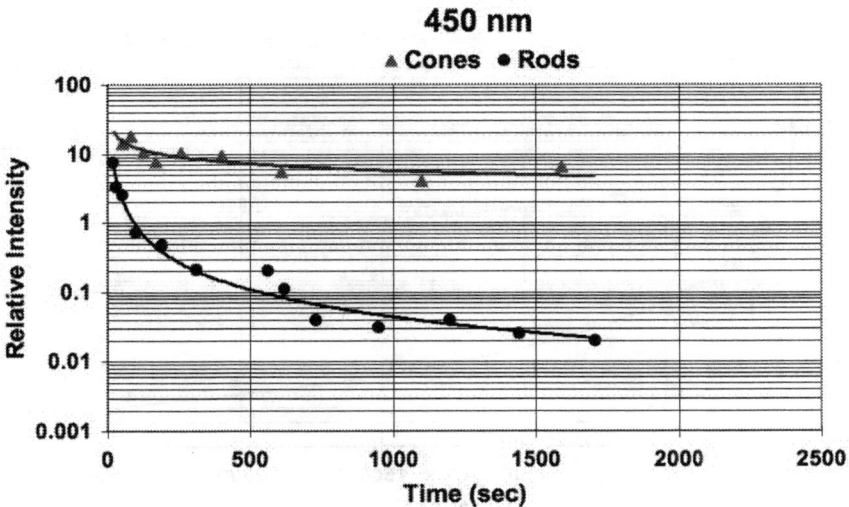

**Figure 31. Visibility thresholds at 450 nm under dark adaptation.**

slit and the trailing shimmer are shown for the 1700 seconds duration of these measurements. The colored slit visibility threshold (labeled as cones in the figure) clearly reaches a minimum level (maximum sensitivity) well before the shimmer visibility. By 1000 seconds (about 17 minutes) the thresholds are still decreasing slowly but the shimmer threshold is two orders of magnitude lower than the color threshold. Apparently, the rods are some one hundred times more sensitive at 450 nm than are the cones. The trend lines shown for each data sequence is a best-fit power law curve.

The sequence for 500 nm was conducted identically and is shown as Figure 32. In this case measurements were carried out at up to 2400 seconds. The shimmer visibility threshold at 1000 seconds is 50 times lower than the color sensitivity. It reaches one hundred times greater sensitivity at 2000 seconds.

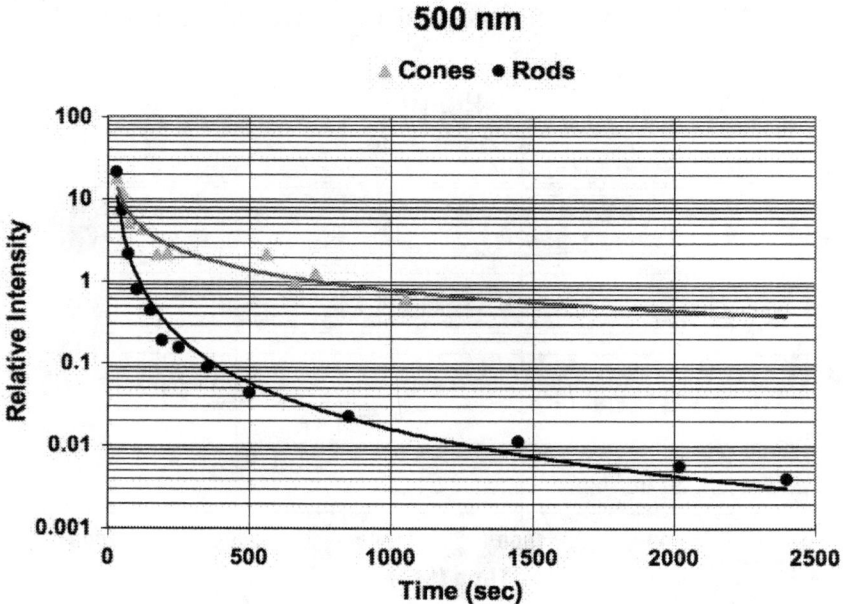

**Figure 32. Visibility thresholds at 500 nm under dark adaptation**

The results for the other two wavelengths tested, 550 nm and 600 nm are shown as Figures 33 and 34, respectively. The sequence for 550 nm was carried out for a total of 2400 seconds. At 1000 seconds; the shimmer threshold was 24 times lower than the color threshold and at 2000 seconds it was 45 times lower.

## 550 nm

### ▲ Cones ● Rods

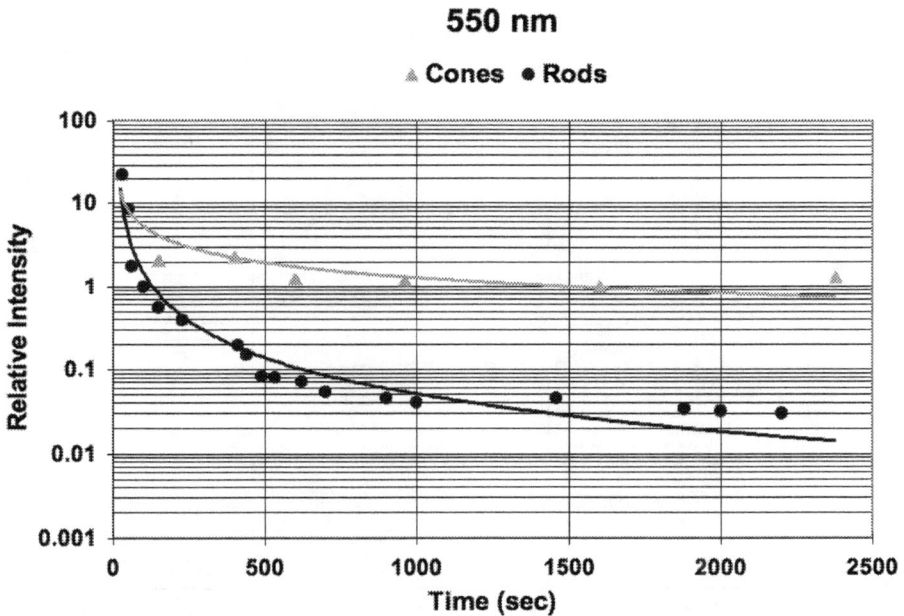

**Figure 33. Visibility thresholds at 550 nm under dark adaptation**

The sequence at 600 nm was carried out to 1700 seconds after which there was little apparent change in threshold. At 1000 seconds the shimmer threshold is only 2.2 times lower than the color threshold. While the apparent separation of thresholds is rather smaller at 600 nm, it is in all cases very easy to observe a different course of dark adaptation for the color perception of the slit and its following shimmer in accord with what one could expect for perception by cones and rods, respectively. In every instance the colors perceived in the leading bar are significantly more saturated than when the bar is stationary. Apparently the simultaneous viewing of the stationary slit

111

by both receptor systems results in the rod perception desaturating the perceived color of the slit. In addition, at 450 and 500 nm, the leading bar is seen as a saturated blue and the following shimmer is white (as it is for any inducing color). Evidently, the rods do not participate in the perception of blue since it clearly desaturates the slit color when stationary.

**600 nm**

**▲ Cones ● Rods**

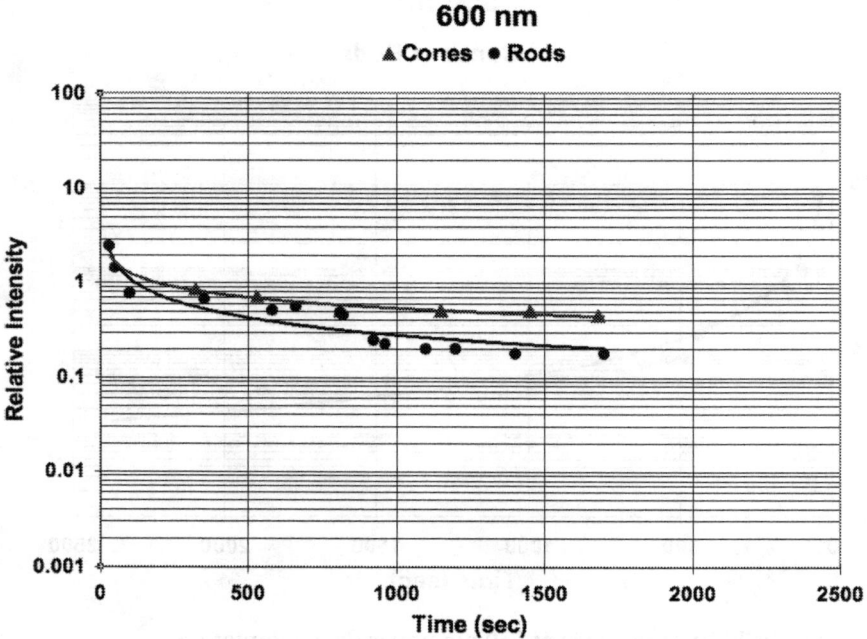

**Figure 34. Visibility thresholds at 600 nm under dark adaptation**

## Experiment III -- Shimmer independence of color

While experiments I and II clearly indicate that the shimmer is due to direct rod perception, further convincing evidence for this can be provided by a third experiment designed to establish the wavelength independence of any temporal latency of the shimmer response. In this experiment we use one bar at a fixed wavelength (500 nm) as a reference and a second bar in which the wavelength was varied. The observer is then asked to position the shimmer of one bar directly in-line with the shimmer induced by the other bar, that is, a shimmer-shimmer match. Keeping the intensities at levels near threshold, the fixed wavelength source is thus used as a reference to examine any differences in temporal latency as the wavelength of the second shimmer source is varied.

An objective measure of the time delay between the passage of the two physical light sources is provided by the timer on the PMT output. For two observers (JM & GC) it was found that the time difference between the appearance of the two shimmer sources was always less than 2.0 milliseconds, regardless of the wavelength to which the varied bar was set. Two milliseconds was, in fact, the effective measurement resolution of our system.

Thus, the shimmer delay was totally independent of the color of the inducing slit image, although the advance time of the colored slit ahead of its following shimmer was not (see below). In summary, these three experiments demonstrate that the shimmer perception has a retinal distribution identical to the anatomical rod distribution, it has a time course of dark adaptation the same as expected for rods, and the time response of the shimmer perception is totally independent of the inducing color. Apparently, the view of a leading colored slit image trailed by its following, whitish shimmer is indeed the separated perception by the cone and rod photoreceptors of the visual system.

**Experiment IV -- Chromatic latency of the cones**
Since the time delay of the shimmer is thus demonstrably independent of the wavelength of the inducing light, the shimmer itself can be used as a constant reference to measure the relative latency of the cone response as a function of wavelength. This effectively provides information on the relative chromatic latency of the cones.

The measurement is done by varying the wavelength of the variable source and keeping its apparent intensity constant at just above threshold. This bar (the relative placement of which is put under observer control) is displaced until its leading part is in line with the

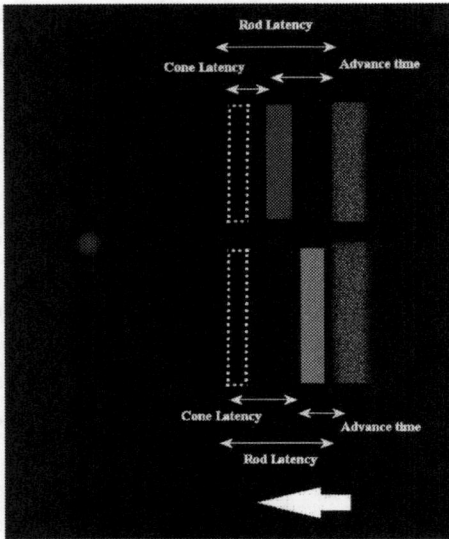

**Figure 35. Advance time definitions**

shimmer of the other bar. The color of the reference slit used in this fashion is set to be the same as that of the test slit, although its exact wavelength does not matter as was shown above.

The geometry of the physical situation – the location of the (unobservable) inducing light source, the leading colored slit, and trailing shimmer - is schematically illustrated in Figure 35. This figure shows what is to be measured (not as set up for the chromatic latency measurements) and illustrates the appearance of displaced red (above) and blue (below) slits advancing from right to left.

The presumptive location of the physical light source is shown as the leading slit outlined in dashes. It is trailed by the cone perception of the colored slit and then the rod perception of the following shimmer. The color with the greatest advance time (preceding the shimmer) is seen to have the shortest latency relative to the physical light source. The figure schematically illustrates the moving slits appearance to the observer (without the outlined physical source, of course) when the two slits are

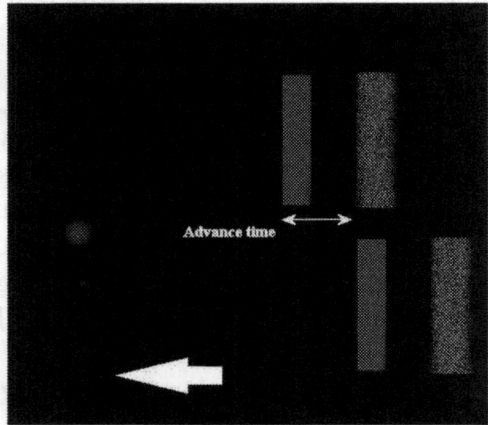

**Figure 36. Geometry of advance time measurement.**

114

initially aligned one above the other when the slits are stationary. When motion is imposed, the shimmer of the two slits maintains the same initial alignment and the colored slit locations are displaced by a wavelength-dependent amount.

The observer view of the actual latency measurements is shown as Figure 36, illustrating two (red) slits displaced by the observer such that the colored slit of one source is exactly in line with the shimmer of the other. As the image of these two displaced slits passes over the photomultiplier entrance slit, the time delay between the leading color edge and the trailing shimmer edge of the separate bars is measured. Since it is the rod-shimmer response that is independent of wavelength, the time difference measured in this way is actually the advance time of the cone system. That is, these measurements are that of the amount of time by which the cone response precedes the constant (with respect to source wavelength) rod response.

Results of these advance time measurements for two observers at 450, 500, 550, 600, and 650 nm are presented in Figure 37. For the range of wavelengths explored, it appears that the temporal response of the cone photoreceptors is a monotonic function of the stimulus wavelength, increasing with wavelength. Again, because of the way the measurements are done, this is not a plot of chromatic latency. Since the larger advance time of the cone response with respect to that of the rods means a smaller latency with respect to the inducing physical stimulus. Thus while the shortest advance time is measured for blue, this actually means that the blue response is the slowest – it is the least advanced from the time of the rod response. Note that the rod response is some 30

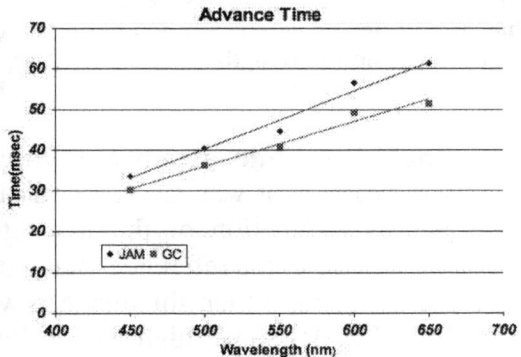

**Figure 37. Cone advance time as a function of wavelength for two observers.**

115

msec more delayed than even the slowest cone response (at the shortest wavelength).

The plotted results are a composite of multiple measurements at each wavelength for the two observers (ten to fifteen separate measurements). The typical standard deviation for each point is ten milliseconds, although the linearity of the plots for the two observers is strikingly evident.

Thus these measurements show that longer wavelengths are perceived first and shorter wavelengths of light have a longer latency in a steadily increasing fashion. These measurements do not determine the absolute cone latency, but only the incremental latency of shorter wavelengths over longer wavelengths. To make this relationship clearer, we plot these measurements as a chromatic latency relative to its (unknown) value at 650 nm as Figure 38. This shows that light at 450 nm takes 20 to 30 milliseconds longer to be perceived than does light at 650 nm and that the lag increases linearly with decreasing wavelength.

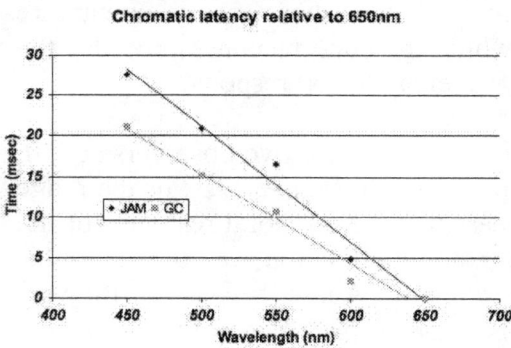

**Figure 38. Chromatic latency relative to that at 650 nm monotonically increasing with decreasing wavelength**

## Experiment V -- Cone latency as a function of intensity

A similar experiment was also carried out to measure the latency of the cones as a function of the intensity of the light (for a fixed wavelength). Here, the reference image intensity is kept constant, and the intensity of the other slit image is varied from the threshold of color visibility to some high upper limit. The observer is then allowed to position the fixed intensity reference source so that its shimmer is in line with the second color (leading) bar of variable intensity. Time differences measured by the passage of the slit

116

images over the photomultiplier are then the advance time of the variable intensity color image with respect to the constant shimmer. Measurements for one observer (GC) at a test wavelength of 550 nm are shown as Figure 39, which plots advance time versus log of relative intensity.

Thus increasing advance time for higher intensities means a shorter latency for a brighter source. For these measurements, we find at 550 nm that increasing the brightness by a factor of a hundred above threshold decreases the latency of the cone response by about 40 milliseconds.

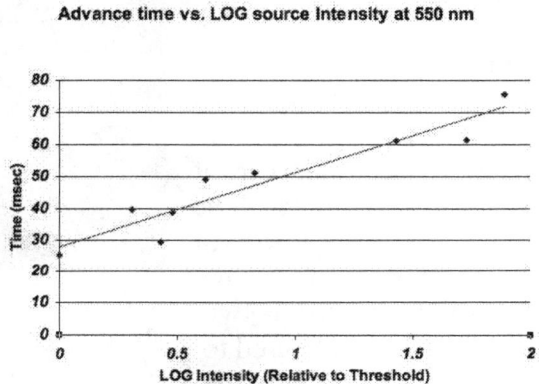

Advance time vs. LOG source intensity at 550 nm

**Figure 39. Advance time variation with intensity at 550 nm.**

## Additional Investigation: Bidwell's Ghost

Although it is quite clear the following shimmer is indeed the direct perception by the rods and there should be little confusion about it as compared to true afterimages, we did make a few observations on the more commonly know effect, Bidwell's ghost, which is apparently a true afterimage as such. The effect is easily distinguished from the rod shimmer since it exhibits the following characteristics:

1) It lags substantially further back than the shimmer (greater time delay),
2) Appears as a distinct, thin bluish bar (thinner than the colored slit or the shimmer),
3) Not readily induced by red light,
4) Most easily induced by green light, and
5) Requires a definite minimum inducing intensity (photopic levels).

The subjective appearance is shown in Figure 40. Again, the white arrow is not part of the scene but only indicates the direction of motion of the slit.

117

We carried out some measurements on the lag of Bidwell's ghost at two wavelengths (500 and 550 nm) where it was relatively easy to invoke the effect. In an arrangement using two identically colored slits, the observer's task was to position the color bar of one slit in direct line above Bidwell's ghost of the other bar. The time interval between the two slits was then measured to provide the lag of the subjective ghost image

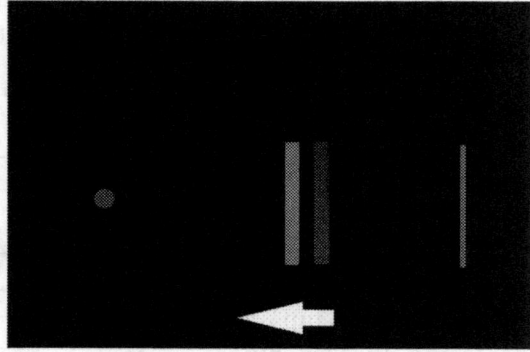

**Figure 40. Subjective appearance of Bidwell's ghost (with fixation light, colored slit, and shimmer.**

behind the perceived color bar. The intensity of the inducing slit was varied, resulting in the data plotted in Figure 41.

As can be seen in the figure, the ghost image lags the colored slit image by about 180 milliseconds at low intensity (the lowest at which is was readily visible). For both inducing wavelengths of light, the lag time decreased monotonically down to about 110 milliseconds at brightness ten times greater. It was difficult to carry these measures out to higher intensities since the ghost image tended to be lost in the glare of the perceived colored slit at any higher intensity.

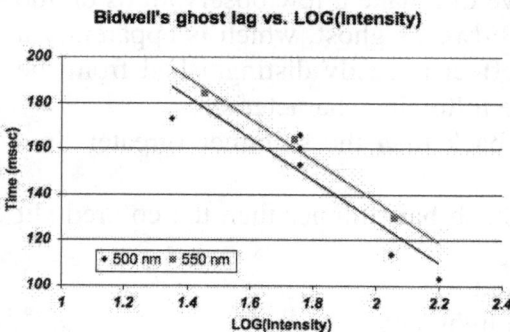

**Figure 41. Time lag of Bidwell's ghost as a function of LOG (Intensity).**

## Summarizing: Rod and Cone Perception

Our measurements on the visual effect we have called the Following Shimmer have clearly shown that the effect is not an afterimage as such, but rather the direct perception of a visual stimulus by the rod photoreceptors. Thus under the conditions of this experiment we have demonstrated the separate and simultaneous perception of a visual stimulus by the cone and rod system of the retina.

That the shimmer is the direct perception by the rods is verified by its absence in the rod-free central area of vision, by the separate dark adaptation curves observed for the colored slit and its shimmer that correspond to what is expected for cones and rods respectively, and finally the demonstration of the complete independence of the time delay of the shimmer with respect to the wavelength of the inducing light.

Given the separate photoreceptor source of color and shimmer, we were able to make use of the shimmer's independence of wavelength to determine the chromatic latency properties of the cones. While it is not possible to determine here the *absolute* chromatic latency of the cones, it was possible to determine that this latency increases monotonically (and apparently linearly) with decreasing wavelength and that at threshold, light of 450 nm wavelength has about a 25-millisecond greater latency than 650 nm light. We also made use of the shimmer's temporal independence to make a few measurements on the effect of intensity on the latency of cone perception. We found that latency does, as expected, decrease with increasing illumination intensity. At 550 nm, the latency decreased by 40 milliseconds with a two-order of magnitude increase in intensity from threshold.

Perhaps one of the greatest values of the present work is the demonstration that this method can be employed as a tool to study basic visual properties and address questions that have long gone unanswered. The experiment does demonstrate that cone and rod activity is simultaneously present in the retina, that rod response is unambiguously white and not involved at all in the perception of blue, that separation of rod perception of a colored object from its cone

119

perception results in a much more saturated color than when both rods and cones participate (i.e., rods tend to desaturate the colors in a retinal image).

Despite various reports in the literature of the blue-blindness of the fovea, our observations revealed no loss of blue perception as the colored slit passed across the fovea – only the loss of the shimmer perception.

In summary this study has, perhaps for the first time, unequivocally elucidated the nature and role of rod perception in color vision. It demonstrates that chromatic latency of the cones is a monotonic function of wavelength, increasing for decreasing wavelength and, hopefully, it has indicated a direction for what could prove to be a powerful new tool for the study of vision. We shall see in the following discussions that the issue of color latency plays a key role in reading the spectral information generated in the proposed cone spectrometer model.

## Dynamics of Visual Perception

A universal aspect of normal vision is the presence of a distinct pattern of involuntary eye movements. Even under rigid fixation when an observer attempts to gaze as steadily as possible towards a fixed point in space, their eyes will nonetheless make small involuntary movements that persist despite all attempts to look in a fixed direction (Ratliff and Riggs, 1950; Riggs and Ratliff, 1951)). As summarized by Sheppard (1968) these involuntary movements consist of three components:

    1.     **Tremor:**  A rapid, irregular oscillation with a frequency range of 30 to 150 Hz, peaking at about 35 Hz with a mean excursion in visual space of about 0.1 minutes of arc.

    2.     **Saccad:**  An abrupt, rapid movement lasting less than 30 msec occurring at intervals ranging from 30 msec to 5 seconds, with excursions in visual space of up to 60 minutes of arc.

3.    ***Drift:***    A slow drifting motion at a rate of about 1 minute of arc/second during the intervals between saccads.

The typical pattern, as illustrated by Ditchburn (1961), is shown in Figure 42 and consists of intermittent rapid flicks with a slow drifting between the flickering saccads and a trembling motion of the gaze during the drifts. The numbered sequence in the diagram labels successive positions as a result of the motions and the tremor is not shown for clarity. The overall effect of these combined motions is to keep the direction of gaze within a region of about 25 minutes of arc in extent around the fixation point. At first sight, it might be natural to assume that these involuntary movements are simply residual instabilities in the ocular muscles' servomechanism controlling the intended line of sight, and thus a hindrance to visual acuity (Ditchburn, 1956). However, this concept is readily disproved by artificially stabilizing the retinal image to remove the involuntary motions in visual space. The result with such stabilized images is that visual acuity is severely impaired, and with sufficiently complete stabilization, the image totally disappears (Riggs, Ratliff, Cornsweet, and Cornsweet, 1953; Ditchburn and Fender, 1955).

When image stabilization is first

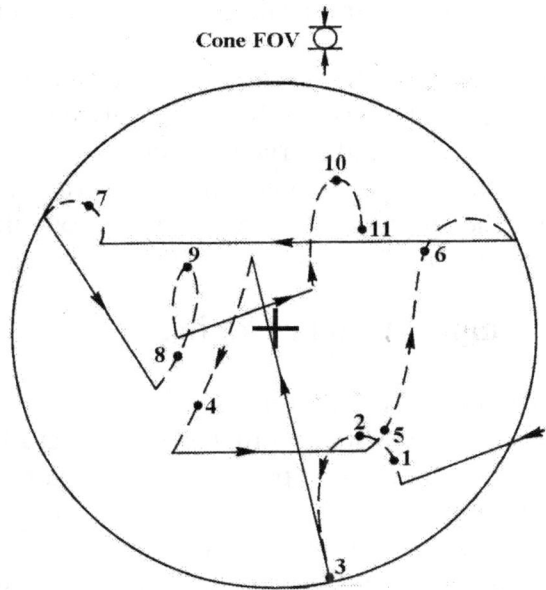

Figure 42. Typical pattern of involuntary eye movements during fixation (Ditchburn, 1961). The numbered sequence shows the successive gaze positions. The circle is 25 minutes of arc in diameter and the approximate field subtended by an individual foveal cone is indicated.

121

turned on, the image is initially seen clearly, but within one to five seconds details disappear, and the field eventually becomes gray or black. After about ten seconds of this, the eye makes a sharp and uncontrollable movement, and unless methods are employed to rigidly cancel image motion, the image is then seen clearly. The image then fades again and the process repeats. The evidence of these kinds of studies is clearly that motion of the image on the retina is necessary to maintain vision.

The contribution of each of the three types of motion to sustain vision has been studied separately. Apparently, it is the saccadic movements that primarily sustain vision with the drifting motion maintaining the general direction of gaze. The drifting itself is not apparently involved in sustaining vision, although the rapid tremor does appear to be involved in sustaining vision in the intersaccadic intervals.

It is notable that under the condition of rigid image stabilization, vision can be restored by temporally modulating the illumination. If a light is flickered (turned on and off) it will be readily visible. The key, of course, is that temporal variation is necessary for vision – whether provided by image motion or intensity modulation – visual perception is evidently a dynamic process.

## Temporal Color Coding

The obvious question for the proposed cone spectrometer operation is: can the spatially ordered wavelength information generated by a cone be detected in principle, and in particular, could it be detected in the retina without obvious violation of current concepts of receptor physiology? There are any number of engineering approaches one could apply in the construction of a physical spectrometer based on this principle. Basically one needs to resolve the positional distribution of light intensity along a linear axis, and the quality of the information one gets will depend on the resolution with which the positional information is read. Note that one could detect either the light distribution inside the cone (where light is differentially attenuated by wavelength) or just outside the cone (where light is differentially shunted out of the cone by wavelength) or both. In any event, one needs to subdivide the cone length into differentially "read" portions; for example, by physically partitioning the cone into regions with separate localized detectors and electrical output lines. A wavelength information-detecting scheme for the retinal cones involving such separately attached readouts seems unlikely, since the observed structure of the cone outer segments does not suggest such a partitioning.

A possible method to read the color information from a cone is to simply convert the positional information into a temporal color code. The basis of this coding is simply that the electrical output of the receptor cannot instantaneously follow variations in the light stimulus, but must exhibit some finite rise time in its response signals. The receptor rise time to follow signal changes will contain coherently-ordered color information, since light-dependent electrical signals generated at the wide, proximal end of the cone have a shorter distance to travel to reach the cone output end than do signal events occurring in the more distal, narrow portion of the cones; thus color-correlated information is spread in time.

An alternative way of picturing how the color-coded temporal order can come about is in terms of electrical capacitance. Red light, predominantly illuminating only the proximal end of the cone, involves signal sources of small spatial extent with presumably smaller time constants (associated with the smaller distributed capacitance) than does blue light, which can illuminate the cone over its full length and involves signal sources with a greater spatial extent (and greater distributed capacitance).

Each time a color border is carried across a cone in the retinal image through the small involuntary eye movements, the time course of the cone output to follow the difference in stimulus is correlated with the color difference at the border. The implications of such an encoding process for the cones are potentially very informative. This concept provides a connecting link between the so-called "blue-blindness" of the central fovea and the tritanopic-like vision observed under conditions of partial or incomplete retinal image constraint. In order to test color discrimination in the very central fovea, very rigid fixation is required since the area to be tested is a small fraction of a degree in extent in the visual field. As we have seen in the discussion of eye movements and retinal image stabilization, such extreme restriction of motion of the field on the retina tends to markedly degrade visual function.

The essence of the proposed model is the production of a specific, coherently ordered, linear pattern of wavelength information in a cone by means of a physical process. To utilize that information does not in itself require the existence of more than one cone photopigment. However, the concept is certainly not incompatible with the presence of more than one. Indeed, from the engineering point of view, the existence of multiple pigments (with absorption maxima well-spaced across the visible spectrum) differentially distributed within the same cone would enhance the efficiency of such a mechanism. There would be considerable operational advantage offered by a long-wavelength absorbing pigment concentrated towards the wide end of the cone and a short-wavelength absorbing pigment concentrated towards the narrow end of the cone. Such a distribution would both enhance the cone sensitivity to longer wavelengths (which are coupled into the cones rather inefficiently) and reduce absorption of shorter wavelength light in the wide portion of the cone thereby reducing some of the ambiguity of the output.

This is, in effect, an amalgamation of the multiple pigment and receptor waveguide concepts, although in a reversal of their traditionally viewed order of importance. Regardless of how the pigments might be distributed in the cones, spectrometer operation would not be hindered by the presence of a long wavelength absorbing pigment in the proximal cone end and a short wavelength absorbing pigment (such as rhodopsin) in the distal portion of the cone. While rhodopsin is usually assumed to be exclusively resident in the rods, the possibility that rhodopsin might be present in the cones is not ruled out on the basis of the objective methods of either

124

extraction or differential bleaching. No photopigment other than rhodopsin has ever been extracted from the primate retina. In the differential bleaching methods, any indications of rhodopsin are categorically ascribed to "contamination" from rod influence.

That rhodopsin is present in the cones at least under some conditions is the implication of the studies on the eye of colorblind humans (monochromats). The existence of a diphasic dark adaptation curve with both the rod and cone components having the action spectrum of rhodopsin in some (so-called typical) monochromats (Alpern, 1974) indicates that whatever photopigment is present in their cones, it has the action spectrum of rhodopsin and the regeneration characteristics of a cone pigment. One possible explanation of these results is that the absorption spectrum alone is a property of the pigment but that its regeneration kinetics are determined in some manner by the structure of the receptor which contains it (Alpern, Lee, and Spivey, 1965; Alpern, 1974).

This also recalls the intriguing speculation of Dartnall (1960) that the cones contain rhodopsin, which has a stable orange photoproduct when present in the cones. This suggestion has some interesting possibilities; it can explain why nothing but rhodopsin can be extracted in the primate retina for which the photopic sensitivity peaks around 540 nm (intermediate between rhodopsin and the photoproduct). If the stability of this photoproduct were somehow associated with the continuous nature of the cone lamella (as opposed to the discrete disk structure in the rods), then the observation that the outer segment membrane in the primate cones is apparently continuous over only its proximal portion, becoming discrete disks more distally (Cohen, 1961), offers a possible mechanism to bring about the differential two-pigment distribution the design engineer would suggest for the cone spectrometer.

There is no question of the importance of the cone photopigments as intermediaries in the transduction of light into electrical signals in the receptor and as determinants of photopic spectral sensitivity. The results of molecular genetic studies clearly indicate the visual system codes for different opsins for different cone pigments. One can question, however, the single-minded effort that has gone into trying to fit spectral sensitivities of proposed cone photopigments to match expectations for the three cone photopigments necessary for the classic Young-Helmholtz model of human color vision. The present model assigns a rather secondary role to the photopigments in the discrimination of color: rather than as a process of continuous intercomparison of static output levels of different receptor types, color vision is portrayed as a dynamic process of reading structure-mediated differences in temporal response characteristics of the cones to light of different wavelength.

In the current context, it is unfortunate that the long-standing emphasis on pigments has dictated rather little curiosity about the physical structure of the receptors themselves and that consequently so little is known about the size and shape of the human retinal cones and their distribution within the retina. More detailed answers to the questions of human color vision need more detailed information on retinal anatomy.

## Temporal Modulation and Spectrometer Coding

Thus far, we have seen that temporal variation is necessary for visual perception and we have suggested that a temporal code would be a natural way to get the spectral information out of a cone. How do these two concepts tie together? On each saccadic eye movement the light input into a given retinal cone will change in synchrony with that movement if the motion carries a color (or intensity) border in the visual field across the cone. As the border is moved across the cone, the illumination within the cone changes instantaneously with that motion, although the electrical response of the detected absorptions will have some latency. We have observed that this latency is greater for blue light (detected at the distal end of the cone) than for red light (detected at the cone's proximal end). What is still required is a means of using this latency to determine the stimulus color.

While we would not directly suggest that the following mechanism is actually the way the cone spectrometer readout occurs in the retina, a conceptually simple scheme for doing this is shown in Figure 43. The distribution of three illuminants along the length of the retinal cone is illustrated along with the time course of a pulse of light. As the light illuminates the cone the output response of the cone is illustrated for the light being either red or blue. In the case of red light the response of the cone output is faster to fall (or rise) than for blue light, which illuminates the entire length of the cone. (Note, by the way, that turning a light stimulus on actually results in a decrease in the cone photocurrent. For the purposes of the current discussion, it does not matter whether the electrical signal rises or falls, but only the rate at which it does so.)

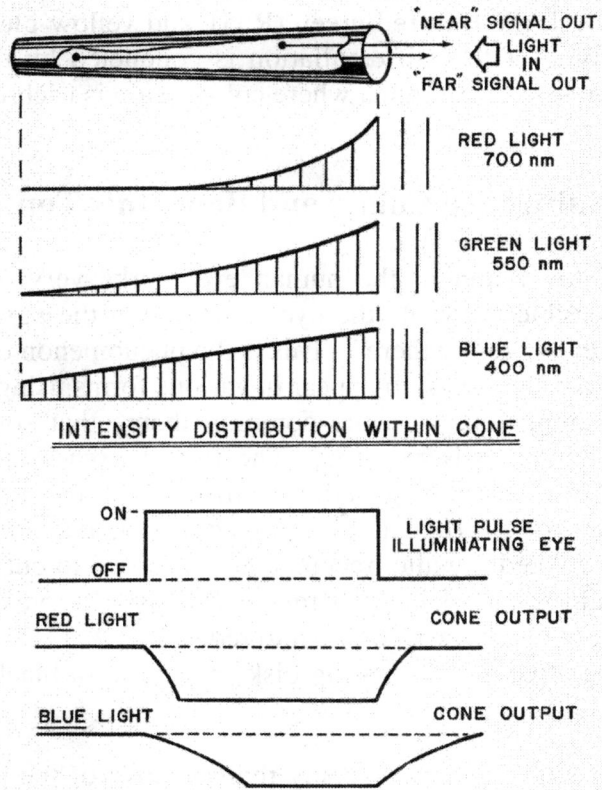

It is a relatively simple task to differentiate the output signals and compare the earlier components of the response to the later components to provide color information. In the case of white (W) light illuminating the

Figure 43. Temporal difference in cone output to a pulse of light

127

cone, the three components of red (R), green (G) and blue (B) will illuminate the cone in a very particular way. The proximal portion of the cone contains R+G+B, the middle portion G+B and the most distal portion B only. On a change in illumination one output can be the detected difference in the proximal portion of the cone minus that in the middle portion of the cone, or (R+G+B)-(G+B) = R. Another output could be the change in the middle portion of the cone less the change in the distal portion or (G+B) – B = G. A third, of course could be simply the change in the most delayed portion from the most distal part of the cone or (B) = B. Once these signals are in hand, total illumination is simply (R+G+B), and yellow (Y) is R+G. This gives the three classic opponent-color dimensions of white-black (W:Bk), red-green (R:G), and yellow-blue (Y:B). Presumably some of this differentiation is conducted by the three bipolar cells present in the retina where color vision is trichromatic.

## Subjective Colors and Benham's Top

Color vision in the human eye works very well indeed, but it is possible to "fool" the eye by invoking the *perception* of colors that are "not really there". This is the phenomenon of subjective color and it can be caused to appear in various forms using only black and white illumination, with the common theme that some form of temporal modulation is required. The first discovery of the subjective color phenomenon is generally attributed to the French monk Bénédict Prevost in 1826 (von Campenhausen, and Schramme, 1995). He oscillated a white, rectangular cardboard to cut across a ray of white light in a darkened chamber. Prevost observed a spectrum of colors from violet through green to red. In 1838, G. T. Fechner rediscovered the effect using rotating disks patterned in black and white (Fechner, 1838).

Around 1894, after many rediscoveries of the phenomena, toymaker C.E. Benham designed a disk (in what was the eighth rediscovery of the effect) sold by Messrs. Newton and Co. that became a popular toy in Victorian England. This disk became known as Benham's top and a typical appearance for the disk is illustrated in Figure 44. Cohen and Gordon (1949) presented a synopsis of research into the phenomena through the early Twentieth Century including the documentation of twelve rediscoveries.

Benham's Top is perhaps the most widely known method for producing subjective colors. It is essentially a disk patterned half in black and half in white with concentric black arcs arrayed on the white sector. When rotated under reasonably bright white light illumination, concentric blur circles at the positions of the sets of arcs are visible. When rotated at speeds of between 5 to 15 Hz, an observer will see these concentric circles as colored in very specific ways. The subjective color effect is often referred to as the Prevost-Fechner-Benham Effect and has been extensively investigated and exhaustively discussed in the technical literature over the last two centuries. Despite the fact that the results of these investigations give clear evidence of the importance of temporal effects on the how the color precept must work in the eye, there is been rather little attention paid to this in standard texts or reviews of human color vision. It is also the case that there is still no accepted accounting for subjective color phenomena and Benham's top still lacks an explanation.

When a disk of the above configuration is rotated in a counter-clockwise direction at around ten revolutions per second, the outer

**Figure 44. Classic version of Benham's Top. When rotated clockwise at about 10 Hz, the innermost arcs appear reddish, the middle arcs appear greenish, and the outermost arcs appear to be bluish to people with normal color vision. On reversing the rotation, the order of color appearance of the arcs also reverses.**

arcs will appear to have a reddish hue with different observers seeing the color with possibly different levels of saturation. The middle arcs will appear to be greenish and the innermost arcs will appear to be bluish. When the direction of rotation is reversed, the order of the evoked color percepts also reverses with the outermost arcs now appearing bluish, the middle still greenish, and the innermost arcs reddish. It is significant that all persons with normal color vision see the exact same sequence, although the degree of saturation in the perceived colors may vary for different observers. It is also notable that it has been reported (Stewart, 1924) that subjective color percepts are evoked in color-blind observers, although "not with the same clarity" as for color normal observers.

Since the time-ordered presentation of achromatic stimuli universally evokes the same ordered perception of color sensation in all color-normal observers, the effect clearly addresses some fundamental property of the human color vision system. What is happening can perhaps be made more apparent if we examine a different form of the Benham's Top configuration. Figure 45 shows a pattern intended to be wrapped around a cylindrical drum that can be rotated about its central axis. The pattern is, again, half-black and half-white with black segments at three positions in the white half. Breaking the length into six equal 60° segments that are viewed as the drum rotates towards the left, the uppermost portion of the drum (segment A) would present successive 60° segments that are sequentially black (B), black (B), black (B), lines (L), white (W), and white (W) with the cycle then repeating. In this sequence the lines of section A will appear to be reddish.

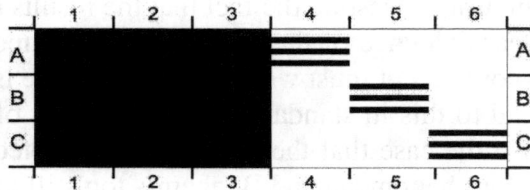

**Figure 45. Rotating drum version of Benham's Top.**

We tabulate the sequences for all the segments A through C under right to left rotation below:

| Segment | Sequence | Evoked Color |
|---------|----------|--------------|
| A | B-B-B-**L**-W-W | Red |
| B | B-B-B-W-**L**-W | Green |
| C | B-B-B-W-W-**L** | Blue |

Reversing the rotation towards the right, the sequence reverses with the color percepts following the same (now reversed) sequence:

| Segment | Sequence | Evoked Color |
|---------|----------|--------------|
| A | B-B-B-W-W-**L** | Blue |
| B | B-B-B-W-**L**-W | Green |
| C | B-B-B-**L**-W-W | Red |

It is thus clear from these tabulated sequences that the time ordering of the lines within the black and white half-cycles is what evokes the color perception. At the risk of belaboring the obvious, we show one more graphic in Figure 46 that is meant to represent the light intensity on the retina as a function of time and the occurrence of the lines within the white hemi-segment. With the end of the black hemi-segment acting as a reset reference, if we start the clock when the white illumination enters the eye then if the line perturbations are at the beginning, the color red is evoked. If the lines are later, the percept is green, and the most delayed line perturbations are seen to evoke the perception of

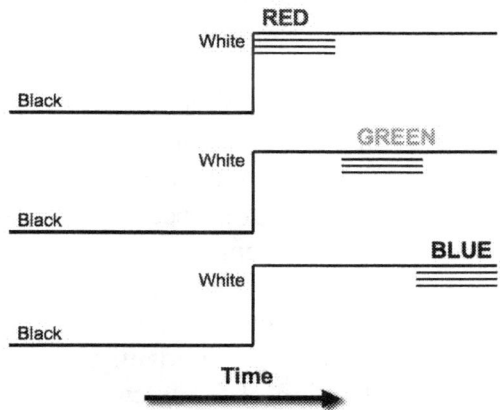

**Figure 46. Time ordered illumination on the retina in Benham's Top and the evoked perception.**

131

blue. Quite clearly this suggests that something in the color vision mechanism associates time constants in the same spectral order: red first, then green, and finally blue with the greatest delay.

As we saw in the previous section on our Following Shimmer experiments, we used the color-independent timing of the rod response as a reference to measure the advance time of color perception by the cones. The key result is that we found that blue perception (at 450 nm) is delayed by 20 to 30 msec compared to red (at 650 nm). How well does this timing compared to that one would infer from the appearance of the spectral colors in the Benham's Top phenomenon?

The comparison does, in fact give remarkably good agreement. Note that color perception in Benham's Top is best evoked for rotation rates of 5 to 15 Hz, depending on the intensity of the illumination. Using an average value of 10 Hz for the rotation rate gives one rotation in 100 msec with each 60° segment occurring within 16.66 msec. This gives the beginning of the evoked blue precept to occur 33.33 msec after the beginning of the red precept and 16.66 msec after its end. This is in very good agreement with our direct measurement of time delay of blue perception relative to red using the following shimmer effect, specifically 20 to 30 msec (Figure 38).

Such a result would hardly seem to be coincidental. Thus, temporally ordering the black and white illumination of Benham's Top with the same time constants observed for human color perception is both a necessary and sufficient condition to evoke the same color perception with achromatic illumination. This result quite clearly demands that at some fundamental level in the visual system, this spectrally-ordered time code is used to encode the basic color information. Evidently, the explanation of subjective color is that modulation of the intensity of white light illuminating the cones is sufficient to evoke the perception of color, so long as the modulation matches the sequence of the properly ordered color code used by the eye to encode color information.

132

# Chapter 7
# Color Blindness

The focus of this volume has been on a discussion of a novel cone spectrometer model for color vision and how it might work in the human eye and thereby explain in a very natural way a number of hitherto puzzling phenomena. But what about how it might not work? It is well known that there are some very specific types and variations of color "blindness". Could the proposed model provide some insight about color defective vision?

## Defective Color Vision

The most common types of color defective vision are X-linked recessive variants of color "blindness" in which function is essentially dichromatic. That is, metameric color matches can be accomplished with only two independent primaries and color vision is essentially dichromatic. Two forms of such dichromatopsia are known; protanopia and deuteranopia. These are different manifestations of so-called red-green color blindness (since the discrimination is essentially only between blues and yellows) and while many of the color confusions are similar in these two forms, there are some distinct differences in their presentation. In addition, both types of red-green color blindness appear to have wide gradations from nearly normal to their extreme, complete forms. These incomplete forms, called anomalous trichromacy, include protanomolous and deuteranomolous variants. Such anomalous trichromats require three color primaries for metameric matches, but the weighting of these matches are different from those of normals (Sheppard, 1968; Sharpe, et al. 1999).

Other versions of color blindness also are known. Much less common than the two forms of red-green color blindness is so-called blue-blindness or tritanopia (Wright, 1952). Color blindness can also be complete so that vision is monochromatic and all color discrimination is absent. There are several variants of monochromatic vision but all such congenital occurrences of monochromacy are extremely rare (Falls, Wolter, and Alpern, 1965). Defects in color vision function can also occur as a result of various ophthalmological and neurological disorders. It can result as well as from extreme exposure to colored lights that upset the adaptational state of vision (Brindley, 1953).

We will primarily limit our discussion here to the most common forms of color blindness, the two forms of red-green color blindness. Standard three-cone models of color vision typically explain color-blindness as being due to the absence of a specific photopigment Graham and Hsia, 1958). Protanopia, for example, is presumed to be explained by the lack of the red pigment and deuteranopia by the lack of the green pigment. While such an explanation seems conceptually straightforward, there are a number of difficulties with this interpretation. For one, the observed spectral sensitivity functions of protans and deutans do not correlate in any simple way with the presumed missing pigments. Accounting for the many versions of protanomolous and deuteranomolous expressions of color blindness (in which red-green color perception is abnormal but not completely lost) requires the presumption of the partial loss (or extra presence) of pigment variations and there is no simple explanation of the observed color vision function in these terms.

Before taking up how defects of the cone spectrometer workings might apply to these common forms of color blindness, we need to examine two key characteristics (and differences) of protan and deutan forms of color blindness. The first is fairly straightforward and has to do with measurements of spectral sensitivity in the two forms. Essentially, protanopic observers have reduced sensitivity at longer wavelengths and deuteranopic observers have normal (or perhaps even enhanced) sensitivity at all wavelengths. The reduced red-sensitivity in protanopia seems to give credence to the concept of a missing red pigment (Hsia and Graham, 1957).

However, the observation that deutan subjects do not have reduced photopic sensitivity compared to normal subjects (Wright and Pitt, 1935: Pitt, 1935) does not fit well with the suggested absence of a green photopigment. Figure 47, a plot reproduced from Graham and Hsia (1958), shows the spectral sensitivity data of Pitt (1935) for dichromats as well as the data of Gibson and Tyndall (1923) for the normal eye. This plot shows the typical procedure of setting the maximum of each sensitivity curve to unity and displaying the results on the same plot. This method does, of course, discard any information on the absolute sensitivities in the different cases.

Hsia and Graham (1957) conducted these sensitivity measurements for dichromats and normals without renormalizing the results so that relative sensitivities could be compared. Their results are shown as Figure 48. Evidently, the maximum absolute sensitivities are similar for both normals and the two types of dichromats, but in different regions of the

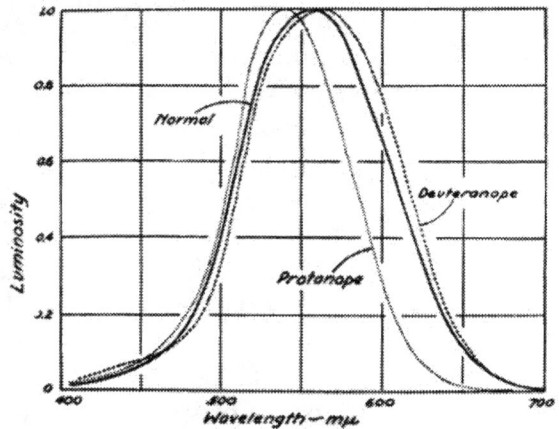

Figure 47. **Spectral sensitivity curves of dichromats (Pitt, 1935) and normal (Gibson and Tyndall, 1923) normalized to unit maximum sensitivity.**

spectrum. Protanopes have absolute spectral sensitivities comparable to normals in the short wavelength portion of the spectrum (below 560 nm or so) and much reduced sensitivity and the longer wavelengths. Deutranopes have absolute spectral sensitivity comparable to normals for the long wavelength portion of the spectrum (above 590 nm or so) and reduced spectral sensitivity for shorter wavelengths (including well into the blue). Hsia and Graham

135

(1957) did note that one of their six deuteranopes did not exhibit this reduced sensitivity at the shorter wavelengths.

More recent results by Kurtenbach, Meierkord, and Kremers (1999) exhibited a similar pattern with protanopes having reduced long wavelength sensitivity and comparable short wavelength sensitivity, although the short wave sensitivity of their protanopes exhibited significant variability.

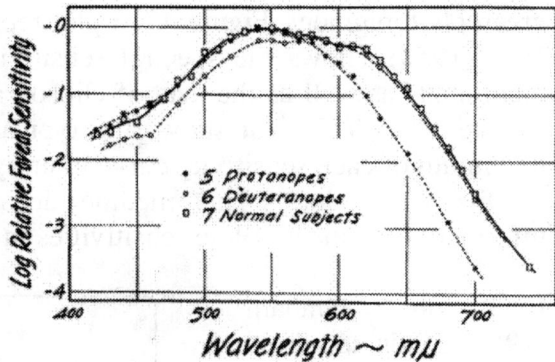

**Figure 48 Spectral sensitivity data of Hsia and Graham (1957)**

Their deuteranopes also showed near normal long wavelength sensitivity with reduced sensitivity at shorter wavelengths, although at the short wavelength end of the spectrum their deuteranopes exhibit substantial variability in spectral sensitivity.

In addition to measures of spectral sensitivity, a second difference in color defective vision compared to normal is their reduced spectral discrimination capability. This deficit essentially defines how these defects may be clinically assessed. A standard diagnostic is to conduct so-called Rayleigh (metameric) matches with a Nagel anomaloscope in which a subject is asked to match a spectral yellow light (589 nm) with a mixture of red (671 nm) and green (535 nm) lights. The only controls the subjects have are to vary the intensity of the yellow light and the ratio of the red and green mixture. Color normal observers tend to use a rather narrow and repeatable range of yellow intensity and green-red mixture ratios to produce a match.

Deuteranopic subjects can use any value of the red-green mixture (since they can not discriminate those wavelengths) and use a yellow intensity similar to that of normals (whatever their relative amount of red or green in the mixture). Protanopic subjects also can use any value of the red-green mixture but require significantly lower luminance of the yellow to match a red setting than a green setting on the anomaloscope (that is, they have reduced red sensitivity).

Subjects with the anomalous forms of color defective vision can discriminate in the red-green region (although in a manner evidently different than normals). Deuteranomolous observers tend to use a repeatable range of red-green mixture more heavily weighted toward green than red compared to normals in order to match the yellow. Protanomolous observers may also have a repeatable range of red-green mixture, but theirs is more heavily weighted towards the red than the green as compared to normals in order to match the yellow.

Curiously, despite the measurably poor red-green discrimination of protan or deutan dichromats, it is not really true that they do not differentiate between red and green. Such protanopes and deuteranopes do, in fact, use the color terms "red" and "green", as well as "blue" and "yellow", and can evidently use them in a plausible way. This has been documented in numerous studies of dichromats (Kalmus, 1965; Boynton and Scheibner, 1967; Scheibner and Boynton, 1968; Jameson and Hurvich, 1978; Wachtler, Dohrmann, and Hertel, 2004). To quote from the paper by Wachtler, et al. (2004) where they first describe some earlier work and then their own approach:

> "That dichromat color percepts are not restricted to a subset of those of trichromats was extensively documented by Scheibner and Boynton (1968). All dichromats tested in this study (3 protanopes and 5 deuteranopes) used "red" and "green" – in addition to "blue" and "yellow" – when tested with monochromatic lights of different wavelengths. The authors suggested that these percepts might be due to residual trichromacy. Molecular evidence for or against this proposal could not be obtained then, due to the lack of

suitable methods to analyze the X-chromosomal opsins at the molecular level.

> "We determined the X-chromosomal opsin gene sequences of dichromats using the polymerase chain reaction (PCR) technique. In two dichromats who were found to have only a single X-chromosomal opsin gene each – either that for the middle-wavelength sensitive M cone or that for the long-wavelength L cone – we confirmed the results of Scheibner and Boynton (1968), using their method of hue scaling of monochromatic lights."

In their experiment, Wachtler, et al. (2004) presented to their subjects either bright or dim narrow-band monochromatic stimuli on bright or dim backgrounds. The procedure used in the experiment is best described in their own words:

> "In each trial, the subject was asked to describe the appearance of the stimulus by giving the relative proportions of primary hues in the stimulus. Subjects were asked to use four color terms "blue", "yellow", "green" and "red" if possible, but were in principle free to use additional terms in case they could not describe their percept with these terms. This situation never occurred."

Wachtler, et al. (2004) went on to document, in careful detail, their procedures for verifying that their dichromats indeed were missing one or the other of the long or middle-wavelength opsin genes. Despite the verifiably absent pigment opsin gene, these dichromats (both protan and deutan) would use the color names with reasonable fidelity (but distinctly different from normal trichromats). Both protans and deutans named "red" and " "yellow" at longer wavelengths and "blue" and "green" at the shorter wavelengths. Interestingly, the deuteranopic observer confined their identification of the "green" and "blue" percept to substantially shorter wavelengths than did the protanopic observer. The protanopic observer did indeed name longer wavelengths "red" (or "yellow") but, unlike normals, never named a wavelength above 530 nm as "green".

Regardless of these details, what is highly significant in the current context is that in the three-cone model of human color vision, if an entire pigment class is totally absent (as is evidently the case here) reasonably accurate color naming is not something that would be possible for these observers, in contradiction to the experimental evidence.

## Defective Cone Spectrometer Tuning

Irrespective of the number, type, and location of multiple photopigments that might or might not be present in various cones, how would cone spectrometer function fail and would it fail in a way that might produce these common forms of color blindness? Critical to the operation of the cone spectrometer is the tuning of its operating range. That is, to discriminate colors in the visible spectrum, the optical waveguide geometry must have the appropriate values (diameters and refractive indices) to disperse the spectrum over its photosensitive length.

As discussed previously in Chapter 5, the ideal operating range for a cone spectrometer is for a range of waveguide parameter, V, values between about 0.8 and 2.5. Values much smaller than 0.8 admit very little illumination within the cone and values greater than 2.405 permit second order mode propagation. We have pointed out that V values just larger than this limit are apparently present in retinal cones for the shortest wavelengths so that violet light simulates the appearance of purple as a mixture of blue and red. In this case, part of the short wavelength light reaches the distal end of the cone in the first order mode (attenuating like blue light) and a portion of the blue light excites the second order mode in the proximal (wider) portion of the cone (attenuating like red light). The range of V values of 0.8 to 2.5 would thus seem to be the optimum range for normal vision.

If the cones are not the right size, however, the operating range would be compromised. Note particularly that we are referring to the optical size of the cones, not just their physical size. It is, in fact, most likely that the cone physical size itself is not the issue, but rather its optical dimensions. This is determined by the difference in refractive index between the interior of the cone and its surround. Given the similarity of construction and content of cones, it is most likely that a mistuning, if it is to occur, would most likely happen through changes in the medium surrounding the cone, the so-called interstitial medium. Recall that the waveguide parameter is given by

$$V = \frac{\pi \cdot d}{\lambda} \sqrt{(n_1^2 - n_2^2)}$$

where d is the receptor diameter, $\lambda$ the free-space wavelength of the incident light and $n_1$ and $n_2$ the refractive indices inside and outside the receptor , respectively. A small change in $n_2$ can effect a marked change in the waveguide parameter, all other things remaining the same. For example, if $n_2$ is only one percent larger than a normal value, than the waveguide parameter will be more than 15 percent (15.3%) smaller than normal (for the same wavelength, cone diameter, and cone refractive index, $n_1$). If $n_2$ is one percent smaller than normal, the waveguide parameter will be more than 18 percent (18.4%) larger than normal. Changes in V as large as this will result in very substantial changes in the relative attenuation along the cone.

While we are assuming here that the most likely way that the cone might be mistuned is through abnormal values of the medium surrounding the cones, we should note that in a study relating molecular genetic measures of photopigments and psychophysical measures of function in protanomolous subjects, Neitz, Neitz, He, and Shevell (1999) suggested that differences in photopigment genetic sequences may regulate the optical density of the cones. This would suggest an alternative means by which genetic mutations coding for photopigments might affect cone function. In this case, differences in photopigments present (or absent in) the cones might alter its refractive index and directly affect the spectroscopic tuning range of the cone.

In any event, for the case of $n_2$ smaller than it should normally be, the range of waveguide parameters will be larger than it should normally be. This mistuned operating range would result in cones that are relatively too large (optically). Since the entrance end of the cone would be too "large" little discrimination would occur there and spectral dispersion would only occur in the more distal end of the cone. The resulting cones would have reduced red-green discrimination (since that discrimination takes place in the proximal end of the cone) and would let in more light overall than a properly tuned cone. That is, these "larger" cones would simulate deuteranopic characteristics. Note that operation of cones significantly larger than normal would have the additional complication of permitting the operation of the second-order $HE_{21}$ mode at wavelengths much longer than in normally-tuned cones (see discussion below).

If $n_2$ is larger than it should normally be, then the range of waveguide parameters will be smaller than normal. These too "small" cones would admit less light than normal and have a reduced sensitivity as compared to normals. Little light would be reaching the distal end of these cones and primarily only blues and yellows would be spectrally dispersed through the proximal portion of the cones. The result would again be red-green color blindness (in a rather different form than the deutan version) with reduced spectral sensitivity – especially in the red end of the spectrum. Retinas with these "smaller" cones would then have defective vision with protanopic characteristics.

The consequence of different cone tunings is schematically illustrated in Figure 49. This shows a 'normal' cone appropriately tuned for the visible spectrum in the middle (same as shown in Figure 12) and a 'too small' cone on the left and a 'too large' cone on the right. The 'too small' cone does not admit any long-wavelength light and would mimic a protan defect. Note that only the proximal portion of this cone provides color dispersion. The 'too large' cone

141

only provides color discrimination in the distal portion of the cone and would mimic a deutan defect.

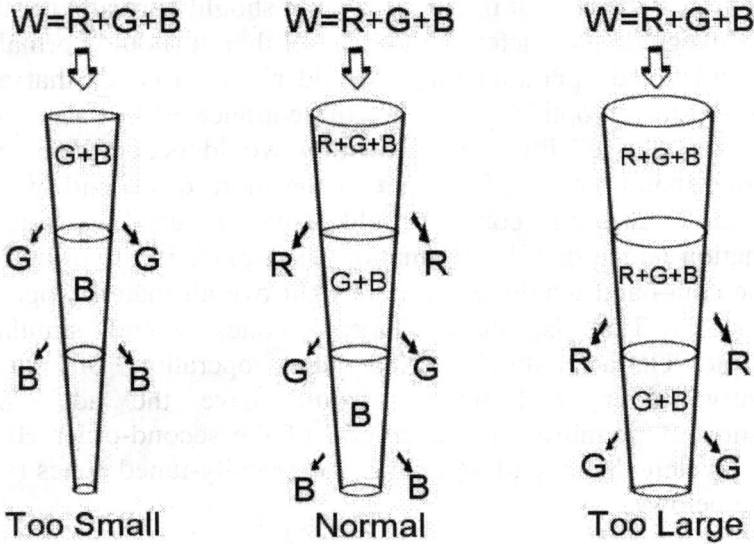

**Figure 49. Light distribution in range of cone 'sizes'.**

Physician L.F. Raymond wrote a series of papers in the early 1970's (Raymond, 1971, 1972, and 1975) in which he described treating patients with various allergies by injections of immunoglobulin E (IgE). He observed that some of the patients that were color blind before these treatments, would subsequently present with normal color vision when examined a few months later. He suggested that this treatment had apparently "cured" the color blindness (unique in the literature as far as I am aware). While there was little detail about the exact type of red-green colorblindness "cured", he did report the successful treatment of some 24 cases.

As a possible reason for these results, Raymond suggested that his treatment for autoimmune problems may have addressed some defect in the production or presence of a photopigment in the subjects' cones. We might suggest a more plausible alternative explanation.

It is notable that IgE is a mucopolysaccharide with some similarities to the mucopolysaccharides known to be present in the interstitial matrix of the retina. Conceivably, the injections of IgE may have altered the composition of the retinal interstitial medium and thereby altered its refractive index in a way that changed the tuning of the color defective retinal cones.

This conjecture is only a supposition about how color blindness "treatment" might be addressed if there is some simple way to adjust the refractive index of the retinal interstitial matrix. In the cone spectrometer model, if one could increase (or decrease) the concentration of mucopolysaccharides in the interstitial medium, the refractive index of the medium would be similarly increased (or decreased). Increasing (or decreasing) the refractive index of the external medium would tune the cone "size" to be smaller (or larger). Since the model suggests that deuteranopic cones are too "large", one would like to increase the refractive index of the medium in their retina to drive the operating point closer to normal. For protanopic cones, one would like to decrease the medium's refractive index to drive the cone tuning towards the normal range.

If mistuning of the cones is indeed involved in red-green color blindness, then there may well be an association between the composition of the interstitial medium in the retina and autoimmune problems in the body. This suggests that it might be very useful to look for any correlation between the presence of autoimmune problems (allergies) and color defective vision.

An interesting consequence of the mistuning hypothesis for color defective vision is that there may be a simple test for demonstrating the mistuned range in color blind observers. Since protanopic cones are presumably tuned too "small", then at no time will the waveguide parameter be as large as 2.405. This means that we would expect protanopic observers to not have any point in the spectrum that would look like the addition of a long wavelength to a shorter wavelength since the second order waveguide mode can never be excited in their cones. On the other hand, since deuteranopic observers presumably have cones that are too large, then this effect of the excitation of the second order mode should occur, but at a longer wavelength than in the violet. For a one percent change in $n_1$, this should occur for a wavelength on the order of 530 nm (instead of the normal 460 nm or so). This prediction should be relatively easy to test.

Another test for this model of color blindness should be possible by testing for the location of maximum color discrimination. Recall that the plot of the derivative of the first-order $HE_{11}$ modal efficiency (Figure 22) shows maximum dispersion to occur at a waveguide parameter value of $V = 1.08$. In normal color vision this should occur at around 580 nm, the unique yellow point at which normal color discrimination is best, about midway along the length of the cone. Again assuming a one percent mismatch in the interstitial refractive index, this best discrimination should occur at around 475 nm for protanopic subjects. While the model would predict best discrimination at around 670 nm for deuteranopic observers, this is probably not observable since there is little differential attenuation in the wider portion of the cone where reds and greens are normally distinguished.

As a final illustration of the possible importance of cone structure and geometry in the current context we consider what has been another, apparently perplexing, observation made on a more extreme form of color blindness, its monochromatic variants. There have been just four reported cases in the literature where histopathology has been performed on the retina of a monochromat (Larsen, 1921; Harrison, Hoefnagel and Hayward, 1960; Falls, et al., 1965; and Glickstein and Heath, 1975). Glickstein and Heath (1975) observed only on the order of 5 to 10% of the normal number of cones in the monochromat retina. Harrison, et al. (1960), also observed an abnormally low number of cones. Falls, et al. (1965), observed reduced numbers of cones outside the fovea, although the fovea itself appeared to contain the normal number of cones. Larsen (1921) found a normal number of cones throughout the retina of his monochromat. However, regardless of the number of cones seen in the various cases, a feature described by all of them was that the cones that were seen were of abnormal size and shape.

The only anomaly described by Larsen (1921) was that of short, abnormally plump cones in the fovea. He observed short, fat cones in the extra-foveal regions, which did not appear to be at all degenerated. Typical descriptions of the cone appearance in the other reports were "abnormal in shape... characteristically much wider than one would expect in a normal human eye" (Glickstein and Heath, 1975), "imperfectly shaped squat cone-like units" (Harrison, et al, 1960).

It is clear that whatever else has gone wrong in these retina, whatever pigments might or might not be missing or whatever defects of neuronal connections might be present, in the context of the present model there is no possibility that these abnormally shaped cones could mediate normal color vision.

# Chapter 8    Future Directions

This work has detailed a number of serious problems with the currently-favored three-cone model of human color vision: the details of the retinal structure suggest that the three "channels" of trichromacy are associated with the post-receptor network and the temporal characteristics of the perception suggest that full-spectral information, in some form, is available at the receptor level. There is no logical way to explain subjective colors with the three-cone model and there are no physical correlates of three cone types in the retina. In addition, the standard model has no obvious explanation of why cones are in fact cones in the first place. There is no place for the observed structure of the color receptors in their functioning. The three-cone model of color vision is so entrenched in the current view that these fundamental difficulties are just simply ignored.

The most important evidence used to buttress the three-cone model is the apparent existence of multiple cone photopigments. It is still the case that, unlike rhodopsin in the rods, cone pigments have yet to be physically extracted from the cones. Regardless, the evidence that the genes do in fact code for multiple cone pigments in the retina is not in dispute. However, the common parlance still confuses multiple pigments and multiple cones. As we have discussed, multiple pigments are not the same thing as multiple cones – multiple pigments are a necessary, but not a sufficient condition for multiple cone types. Multiple pigments might, for example, be used to enhance the operational efficiency of some other means of discriminating color.

I have proposed exactly this for the spectroscopic model of a retinal cone where dimensional change (conical taper) mediates the spatial sorting of light according to its wavelength along the length of the cone. In this case, multiple pigments can enhance the efficiency of this basic mechanism through the presence of a long-wavelength absorbing pigment in the wide, proximal end of the cone and a short-wavelength absorbing pigment in the narrow, distal end of the cone.

A survey of the meager data available on human retinal cone shape, dimensions and refractive index indicates that the receptors are both tapered (including the foveal receptors) and small enough that they should explicitly exhibit the color sorting process proposed. We have demonstrated and photographed this spectral sorting of color by a cone in small, tapered optical fibers. The observed spectral dispersion is just that pattern predicted by fundamental waveguide physics and is what we predict to be present in the retinal cones.

Recall that in the three-cone model of human color vision, each receptor type absorbs a given colored light with a probability determined by its photopigment absorption curve. Each cone type responds indifferently to the color of the incident light, except that its response amplitude is greater for a color for which it has a greater absorption probability. As a consequence of this basic property of the mechanism, any information about the spectral content of the incident light is discarded and unavailable at the cone output. It is only by inter-comparison of the output of the various cone types that color information can be deduced. For a three-cone model, then, a metameric match between a pure spectral yellow and a mixed green and red that appears yellow when viewed under static conditions must remain a match even under dynamic presentation of the two ways of producing yellow. This fundamental prediction of the three-cone model of human color vision is flatly contradicted by experimental results.

This breakdown of statically established metameric matches under dynamic conditions – as first described by Ives (1918) – is easily reproduced. I described our experimental observations using a moving slit apparatus that we assembled for this purpose. I further described the results we got with the apparatus to directly observe the separation of rod and cone perception and its use to measure the chromatic latency of cone perception. We found a linear increase in the latency of color perception with decreasing wavelength – an impossible result in a three-cone mechanism.

The operation of the proposed cone spectrometer model can be simply

148

summarized as a process in which waveguide mode cutoff disperses light along the length of the cone outer segment according to its wavelength and modulations in light illuminating the cones, through saccadic movements, converts the positional color information into a temporal code. In detail, the steps involved in the operation of the cone spectrometer model proposed here can be itemized as follows:

- Light is coupled into the photosensitive outer segments of the cones at its broad entrance end,
- Due to its dimensions and refractive indices the cones have waveguide parameters near cutoff for the lowest-order waveguide mode, the HE11,
- For the shortest visible wavelengths, in the violet, the next order (HE21) mode can also be excited at the widest end of the cone leading to a pattern simulating a mixture of red and blue,
- As light propagates along the length of the cone in the direction of decreasing diameter, the waveguide parameter decreases for a given incident wavelength so that the power efficiency of waveguide propagation decreases resulting in a progressive shifting of the guided energy from within the cone to just outside it as the exponentially-damped evanescent wave,
- This dispersive effect is differential with optical wavelength so that the longest wavelengths are shunted out of the cone interior the most rapidly,
- In a properly tuned cone, this gives the resulting illumination pattern in the cone interior of all wavelengths illuminating the wide, entrance end of the cone, successively shorter wavelengths being shunted out of the cone as the light propagates in the direction of decreasing cone diameter until only the shortest wavelength light remains to illuminate the narrowest, distal tip of the cone,
- Pigment absorption along the length of the cones initiates the phototransduction process where ion channels are closed along the cone wall, reducing the normally present dark current,
- This photocurrent reduction is localized to a longitudinal position along the cone outer segment to within one micrometer or less of the position of a given photoabsorption event,

- The photocurrent reduction is transmitted to the cone output end, at its junction with the bipolar cells, with a time delay correlated with its distance down the length of the cone outer segment so that the most distant absorption events are the most delayed in reaching the cone output,

- This process results in a correlation between the time delay in cone output signal and the position of an absorption event along the cone length and consequently with the wavelength of the incident light,

- The saccadic movements of the eye provides a natural time reference to interpret this correlation since at each shift of the retinal image there is a corresponding abrupt shift in the illumination within a cone as any border in color or illumination passes over any given cone (in lieu of such motion during retinal image stabilization, any imposed temporal modulation of color or intensity will suffice to provide the time reference),

- It is the pattern of these absorption events that can be interpreted to infer the color of the light illuminating the cone since all light illuminates the entrance end of the cone outer segment although only the shortest wavelength light can illuminate the distal end of the cone,

- This pattern of photocurrent fall (or rise) times with illumination increase (or decrease) must be differentiated by the retinal circuitry to interpret color, presumably in a standard opponent colors form of red-green, blue-yellow, and light-dark,

- There must, in addition, be some form of comparison and differentiation among the signals from many cones to interpret and discount the color cast of the general, overall illumination of a scene in order to support the remarkable color constancy ability of the eye to properly interpret colors despite wide variation in general illumination.

With no specific assumptions about the number and type of cone photopigments that might or might not be present in the eye and only minimal assumptions about how the positionally correlated color information is converted into a temporal code, the basic concept of the spectroscopic cone model offers a framework in which a broad range of information about human color perception can be understood, including:

- The conical shape and small size of the color receptors,
- The correlation between cone length and color resolution,
- The dynamic breakdown of static metameric matches,
- That spectral information is not lost at the receptor level,
- That any color, including white, is perceived by a single cone,
- The special status and high colorimetric purity of blue light,
- The similarity of violet and purple,
- The abrupt minimum in color discrimination at around 440 nm,
- Some of the adaptational properties of human color vision
- The predominant red-shift characteristic of the SC II effect,
- The general colorimetric purity characteristics of color vision,
- The general color discrimination characteristics of color vision
- The ordered chromatic latency of color vision,
- Explanation of subjective colors in terms of chromatic latency, and
- A pigment-independent way of understanding the common forms of color blindness that might yet lead to some therapeutic approaches.

Notwithstanding the obvious utility of the model in helping to understand these phenomena, there is still a good deal that has not been addressed in this volume and much work yet to do to fully understand human color vision. For example:

- How exactly is the temporal coding of the spectral information converted into the three channels of normal color vision?
- How are multiple pigments distributed in the cones and what is their role in enhancing the basic discrimination mechanism?
- There are a number of effects and characteristics of color vision that will need to be addressed in detail in the context of the current model and these are clearly fertile ground for further research. These properties include such features as the detailed shape of the hue and saturation curves of color vision, the

observed improvement in color discrimination in the blue with the addition of desaturating white light (Tyndall Effect), the complex changes in apparent hue with intensity (Bezold-Brücke Effect), the detailed explanation of the entire range of the various forms of color defective vision, and a more detailed study of a possible subpopulation of "blue cones".

- How is color constancy implemented with this mechanism? Clearly, there must be comparisons among the cones of the general illuminant on the retina to properly interpret the color signals from individual cones despite wide variations in source illumination intensity and color cast. How is this effected?
- What role does cone spectrometer operation play in the color vision of other species? Well-defined cone pigments with sensibly positioned spectral sensitivities have been identified in multiple cone classes in goldfish, for example. Many species employ colored oil droplets between the inner and outer segments of their cones to act as color filters. Not withstanding the existence of these alternative mechanisms, the retinal cones in these other species are still conical, so is the spectroscopic action of the cone taper ignored or is it taken advantage of in some way?
- Can we modify the refractive index of the interstitial matrix in the retina to address color vision deficiencies? This may have already been done serendipitously for possible remission of color defective vision. Is there some way to do this in a more systematic way that might offer a long term improvement in the quality of vision for certain types of color blindness?

Given the current near-universal acceptance of the three-cone model as the explanation of human color vision, there will certainly be significant resistance to easy acceptance of the proposed cone spectrometer model. Given the radically different nature of the proposed mechanism, that is precisely as it should be. Scientific advances depend on the explanatory and predictive power of new models and the repeatability of experimental results. The proposed model and experimental results presented here need to stand or fall on independent investigation leading to confirmation or contradiction.

I do hope that the current effort to point out some of the fundamental flaws and general explanatory weaknesses of the standard model on the one hand and, on the other, to present the evidence supporting the proposed cone spectrometer model and its explanatory capabilities will, in time, find color vision researchers willing to fairly explore this approach. The view should be that the issues set out in this volume are a potentially useful beginning. The ultimate goal should be nothing less than a more complete understanding of human color vision.

*John A. Medeiros*

## CONE SHAPE AND COLOR VISION:
UNIFICATION OF STRUCTURE AND PERCEPTION

# *APPENDIX*

# Dielectric Cylinder Waveguide Propagation

## *General Solution*

The spectral dispersion characteristics of the tapered dielectric cylinder follow from an analysis of the electrodynamical problem. Much of the effort in this analysis should be applied to an understanding of the simple, uniform dielectric cylinder since problems involving non-uniform geometries can be mathematically synthesized from the uniform cylinder solution. That is, the full spectrum of eigenvalue solutions of the uniform dielectric cylinder problem constitute a complete orthogonal basis set for a series expansion representation of the solution for more complex geometries. In the special case of very small and only slightly tapered cylinders, nearly all the information needed for solution may be taken directly from the uniform cylinder solution.

There exists a considerable body of literature on the dielectric cylinder waveguide; this is reviewed and many details of the problem are summarized in, for example, Kapany and Burke (1972), Marcuse (1974), and Snyder and Love (1983). Here we present only an outline of these computational details and a summary of the main results. This analysis is used as the basis of a simple computational program to solve the eigenvalue problem and provide a basis for the numerical examination of the waveguide properties.

155

The geometry of interest is shown in Figure 50 where we depict a uniform, open, bounded dielectric cylinder of diameter d (radius a = d/2) and refractive index $n_1$ embedded in an infinite medium of refractive index $n_2$. The permittivity and magnetic permeability of the cylinder and its surround are, respectively $\varepsilon_1$, $\mu_1$ and $\varepsilon_2$, $\mu_2$. The magnetic permeability of all dielectric materials of practical interest may be taken to be that of free space. We thus use $\mu_1 = \mu_2 = \mu$. The z-axis is set to be coincident with the cylinder axis of symmetry and a plane wave of wavelength $\lambda$ is incident with wave vector $\bar{k}$ (magnitude $|\bar{k}| = 2\pi/\lambda$) at an angle $\theta$ with respect to the z-axis.

This is, of course, a highly idealized geometry. In real physical cases of interest we will not have infinitely long nor completely isolated structures. Nonetheless, this ideal geometry is of practical importance since the perturbing effects at the cylinder ends and of adjacent structures are of negligible importance beyond a distance of a few wavelengths. Aside from considerations of how particular propagation modes are excited and coupled into the cylinder for given conditions at its entrance end, the infinite cylinder representation is an accurate one.

The problem consists of determining the electric and magnetic field vectors ($\vec{E}$ and $\vec{H}$, respectively) that are the solution of the homogenous wave equation

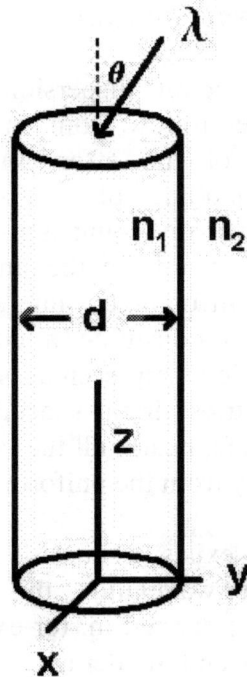

**Figure 50.** **Parameters of the infinite dielectric cylinder.**

156

$$\left[\nabla^2 - \varepsilon\mu\left(\partial^2 / \partial t^2\right)\right]\begin{Bmatrix} \vec{E} \\ \vec{H} \end{Bmatrix} = 0. \qquad (1)$$

where $\varepsilon$ and $\mu$ are the dielectric constant and magnetic permeability, respectively, of the medium. A particular solution is obtained by applying the appropriate boundary conditions.

For the dielectric rod the tangential components of the electric and magnetic fields of the equation are required to be continuous across the rod-surround interface. We are, in addition, interested in those solutions representing local confinement to the rod structure (guided waves). We thus match solutions at the boundary for fields that are zero at infinite radial distance (non-radiative propagation) with fields that are finite within the waveguide. These are the elementary solutions of the form

$$\begin{Bmatrix} \vec{E}(x,y,z,t) \\ \vec{H}(x,y,z,t) \end{Bmatrix} = \begin{Bmatrix} \vec{F}(x,y,z,t) \\ \vec{G}(x,y,z,t) \end{Bmatrix} \exp\left(i\omega t - ihz\right). \quad (2)$$

Here, h is the z-component of the wave vector (namely, $k\cos\theta$).

The general solution procedure consists of substituting this form for the fields in the wave equation (1) and then solving for the eigenvectors $\vec{F}$ and $\vec{G}$ and the eigenvalues h. The eigenvalue equation is

$$\left(\nabla_t^2 + \beta^2\right)\begin{Bmatrix} \vec{F}(x,y,z,t) \\ \vec{G}(x,y,z,t) \end{Bmatrix} = 0 \qquad (3)$$

where

$$\beta^2 = \omega^2\varepsilon\mu - h^2 = k^2 - h^2 \qquad (4)$$

157

is the propagation constant of the medium and the transverse Laplacian operator is

$$\nabla_t^2 = \frac{\partial^2}{\partial x^2} + \frac{\partial^2}{\partial y^2} = \frac{\partial^2}{\partial \rho^2} + \frac{1}{\rho}\frac{\partial}{\partial \rho} + \frac{1}{\rho^2}\frac{\partial^2}{\partial \phi^2}$$

where $\rho$ and $\phi$ are the polar coordinates with $x = \rho\cos\phi$ and $y = \rho\sin\phi$ and the inverse relations $\rho = (x^2 + y^2)^{\frac{1}{2}}$ and $\phi = \tan^{-1}(x/y)$.

We will use the subscript 1 to refer to the values of the media constants inside the guide and the subscript 2 for the constants outside the guide. Thus, for example, we will have inside the guide the values $\beta_1$, $k_1$, $n_1$, $\varepsilon_1$, $\mu_1$, and $h_1$. Note that for common dielectric media of interest, $\mu_1 = \mu_2 = \mu$ and $h_1 = h_2$.

Since the electric and magnetic fields are also solutions of the first order Maxwell equations:

$$\nabla \times \vec{E} = -\mu\frac{\partial \vec{H}}{\partial t}$$

$$\nabla \times \vec{H} = \varepsilon\frac{\partial \vec{E}}{\partial t} \tag{5}$$

then, substituting the forms of $\vec{F}$ and $\vec{G}$ and decomposing the eigenvectors into longitudinal ($\vec{F}_z$ and $\vec{G}_z$) and transverse ($\vec{F}_t$ and $\vec{G}_t$) components we get, after suitable manipulation (c.f., Kapany and Burke, 1972) the conditions

$$\beta^2\vec{F}_t = -i\omega\mu\nabla_t \times \vec{G}_z - ih\nabla_t\vec{F}_z$$

$$\beta^2\vec{G}_t = -i\omega\varepsilon\nabla_t \times \vec{F}_z - ih\nabla_t\vec{G}_z \tag{6}$$

Thus if the wave equation (3) is solved for the longitudinal components, then the transverse components can be generated through equation (6). Since the fields must be periodic in $\phi$ we use a

158

trial solution of the form $F_z = f(\rho)\exp(in\phi)$ in the wave equation with n an integer. Separation of variables is then possible and the radial equation for $f(\rho)$ becomes Bessel's differential equation,

$$\frac{d^2 f}{d\rho^2} + \frac{1}{\rho}\frac{df}{d\rho} + \left(\beta^2 - \frac{n}{\rho^2}\right)f = 0. \qquad (7)$$

Thus the solutions for $F_z$ are of the form

$$F_z(\rho, \phi, h, \omega) = AZ_n(\beta\rho)\exp(in\phi) \qquad (8)$$

Where $Z_n(\beta\rho)$ is a Bessel function of order n.

Thus far this solution is quite general and a particular problem is solved by determining those particular forms of trial solutions that satisfy the applicable boundary conditions (if such can indeed be found). Before carrying out this procedure for the dielectric cylinder, we note that while cylindrical coordinates are indeed the natural representation for this problem, there is no preferred direction along which to orient the x-axis (where $\phi = 0$). This azimuthal symmetry suggests circulating wave solutions to the wave equation with elliptical polarization in the transverse plane. Anticipating this result, Kapany and Burke (1972) suggest the unconventional, but very useful, representation for the field as complex transverse field components in the form $E_+$, $E_-$, $E_z$ and $H_+$, $H_-$, and $H_z$ where, for example,

$$2E_\pm = E_x \pm iE_y \qquad (9)$$

and

$$\bar{E}_\pm = E_\pm(\hat{x} \pm i\hat{y})e^{i\omega t - ihz} \qquad (10)$$

where E± is the complex-valued scalar with amplitude $\left|\bar{E}_\pm\right|$ and phase $\Delta\pm$ and similar equations obtain for the magnetic field components. $\bar{E}_+$ represents a vector (amplitude $\left|\bar{E}_+\right|$) rotating counter-clockwise with respect to the axis perpendicular to the x and y-axes (the z-axis). $\bar{E}_+$ is a circulating wave of positive helicity. Similarly, $\bar{E}_-$ is a clockwise rotating vector representing a circulating wave of negative helicity. Thus the $\bar{E}_+$ and $\bar{E}_-$ are, respectively, the left circularly polarized and right circularly polarized field components. $\bar{E}_\pm$ and $\bar{H}_\pm$ are orthogonal in the power sense with

$$\bar{E}_\pm \times \bar{H}_\pm^* \propto (\hat{x} \pm i\hat{y}) \times (\hat{x} \mp i\hat{y}) = 0$$

where the asterisk denotes complex conjugation. In terms of the more conventional field vector representations, we have

$$2\bar{E}_\pm = \bar{E}_x \pm i\bar{E}_y, \bar{E}_x = \bar{E}_+ + \bar{E}_-, i\bar{E}_y = \bar{E}_+ - \bar{E}_- \quad (11a)$$

$$\bar{E}_\rho \pm i\bar{E}_\phi = e^{\pm i\phi}(\bar{E}_x \pm i\bar{E}_y) = 2e^{\mp i\phi}\bar{E}_\pm. \quad (11b)$$

Note that in this definition, the circulating fields are not normalized and the transverse gradient operator is thus

$$\nabla_\pm = \frac{1}{2}e^{\pm i\phi}[(\partial/\partial\rho) \pm (i/\rho)(\partial/\partial\phi)] = \frac{1}{2}\nabla_t. \quad (12)$$

So for field components of the form $F_n = AZ_n(\beta\rho)e^{in\phi}$ and $F_{n\pm1} = AZ_{n\pm1}(\beta\rho)e^{i(n\pm1)\phi}$ as in equation (8), we have

$$\nabla_\pm \bar{F}_n = \mp\frac{1}{2}\beta\bar{F}_{n\pm1}$$

$$\nabla_\pm \nabla_\mp \bar{F}_n = -\frac{1}{4}\beta^2\bar{F}_n$$

$$(13)$$

The second-order equations (13) are equivalent to the wave equation (3). In this circulating vector field representation, the results of the procedure outlined above give the eigenvector solutions:

$$\vec{E}_{\pm} = \pm(1 \pm \alpha)A\vec{F}_{n\pm 1}$$

$$\vec{E}_{z} = (2\beta/ih)A\vec{F}_{n}$$

$$\vec{H}_{\pm} = (ih/\omega\mu)\big[(k^2/h^2) \pm \alpha\big]A\vec{F}_{n\pm 1} \qquad (14)$$

$$\vec{H}_{z} = (2\beta/\omega\mu)\alpha A\vec{F}_{n}$$

$$\vec{F}_{n} = \vec{Z}_{n}(\beta\rho)\exp(in\phi + i\omega t - ihz)$$

These equations are sufficiently general to describe the fields both inside and outside the cylinder although it is yet to be shown that the solutions can, in fact, be matched across the cylinder wall. The constants A, h, $\beta$ and $\alpha$ can not be assumed equal in both the cylinder interior and the surround. The dielectric constants $\varepsilon_1$ and $\varepsilon_2$ are different in the two regions and thus $k_{12} = \omega_2\varepsilon_1\mu$ and $k_{22} = \omega_2\varepsilon_2\mu$ are different. The subscripts 1 and 2 are used to identify parameters of the cylinder and surround, respectively.

The boundary conditions required for the dielectric cylinder give the following constraints:

The fields must be periodic in $\phi$ so that n is restricted to integer (or zero) values.
The fields must be finite on the cylinder axis ($\rho = 0$) so that in medium 1, $Z_n(\beta_1\rho) = J_n(\beta_1\rho)$ where $J_n$ is the Bessel function of the first kind of order n.

Outside the guide, two possibilities may be identified; the fields are either localized to the vicinity of the cylinder (guided modes) or else radiate away from the cylinder. In the first case the external fields are evanescent with amplitude decreasing to zero with increasing radial distance from the cylinder. The $Z_n(\beta_2\rho)$ satisfying this condition are the modified Bessel functions of the second kind, $K_n(\beta_1\rho)$. In the second case when the fields radiate energy away from the cylinder, the Sommerfield radiation condition must be satisfied for $\rho \rightarrow \infty$ and the appropriate Bessel functions are the Hankel functions of the second kind $H_n(2)(\beta_1\rho)$ (Kapany and Burke, 1972). The Bessel functions outside the guide are interchangeable through $H_n(2)(-iq) = (2/\pi)e^{(n+1)\pi i/2}K_n(q)$. The Hankel functions obey the same recursion relations as do the $J_n$.

1.    The tangential components of the fields must be continuous across the cylinder wall (at $\rho = d/2$). For the longitudinal components we require $E_{1z}(\rho=d/2) = E_{2z}(\rho=d/2)$ and $H_{1z}(\rho=d/2) = H_{2z}(\rho=d/2)$. These give the propagation constants to be related as:

$$h_1 = h_2 = h$$
$$\alpha_1 = \alpha_2 = \alpha \qquad (15)$$
$$A_2 = \frac{uJ_n(u)}{\upsilon H_n(\upsilon)}A_1$$

using the definitions, $u = \beta_1 d/2$ and $v = \beta_2 d/2 = iq$. The constant $A_1 = A$ is arbitrary with $AA^*$ proportional to the modal power.

The matching of the azimuthal components of $\bar{E}$ and $\bar{H}$ lead to the characteristic equations for the dielectric cylinder waveguide. The equations, after suitable algebraic manipulation can be written in two forms; one form specifies the mode constant $\alpha$ and another form that gives the eigenvalue equation. For $\alpha$, which effectively specifies the polarization characteristics of the mode, we have

$$\alpha = n\left[\frac{1}{u^2} + \frac{1}{q^2}\right]\left[\frac{J_n'(u)}{uJ_n(u)} + \frac{K_n'(q)}{qK_n(q)}\right]^{-1}. \qquad (16)$$

The eigenvalue equations are

$$\frac{J_{n\pm1}(u)}{uJ_n(u)} = \frac{n}{u^2} \pm (1-\delta/2)\left(\frac{K_{n-1}(q)}{qK_n(q)} - \frac{n}{q^2}\right) - \left[\frac{\delta}{4}\left(\frac{K_{n-1}(q)}{qK_n(q)} + \frac{n}{q^2}\right)^2 + n^2\left(\frac{1}{u^2} + \frac{1}{q^2}\right)\left(\frac{1}{u^2} + \frac{1-\delta}{q^2}\right)\right]^{1/2} \quad (17)$$

where $\delta$, the dielectric difference, is defined by

162

$$\delta = 1 - \varepsilon_2/\varepsilon_1. \qquad (18)$$

The eigenvalues $u$ and $q$ are related to the physical parameters of the guide. The dimensionless parameter $V$ is defined by $V^2 = u^2 + q^2$. This may be related to the guide physical parameters using the definitions: $u = \beta_1 d/2$ and $v = \beta_2 d/2 = -iq$ ($q$ real) and equation (4) for the propagation constants in each medium. The result is

$$V^2 = u^2 + q^2 = \left(k_1^2 - k_2^2\right)\frac{d^2}{4} = \left(\frac{\pi d}{\lambda}\right)^2 \left(n_1^2 - n_2^2\right). (19)$$

The two equations (17) define two mode classifications; with the choice of the upper (positive) sign, it specifies the eigenvalue solution for the so-called $EH_{nm}$ modes (with eigenvalue designated as $u_{nm}^{EH}$) and with the choice of the lower (negative) sign, the $HE_{nm}$ modes (with eigenvalue designated as $u_{nm}^{HE}$). For each value of n, that value of $u$ (and consequently, $q = (V^2 - u^2)^{1/2}$) for which the transcendental functions on either side of the equals sign in equation (17) equate (intersect) must be found. For these transcendental equations this, in general, can only be done numerically (or graphically). For each value of $n$, successive roots of equation (17) are labeled by the subscript $m$. The eigenvalue bounds are found to be

$$[m^{th} \text{ non-zero root of } J_{n-2}(u) = 0] < u_{nm}^{HE}$$
$$< [m^{th} \text{ non-zero root of } J_{n-1}(u) = 0]$$

$$[m^{th} \text{ non-zero root of } J_n(u) = 0] < u_{nm}^{EH}$$
$$< [m^{th} \text{ non-zero root of } J_{n+1}(u) = 0]$$

Note that there is only one solution for values of $u$ less than 2.405. This is for the $HE_{11}$ mode where $J_{-1}(u) = 0$ and $J_0(u) = 0$ give the bounds $0 < u_{nm}^{HE} < 2.405$. The two mode classifications physically represent two different ways of combining orbital angular momentum (from the phase factor) and spin angular momentum (polarization). For the $HE_{nm}$ and $EH_{nm}$ classifications, the total field angular momentum (proportional to $-n$) represents, respectively, additive and subtractive combination of the angular momentum components.

In summary, then, we have the following prescription for determining the fields in a cylindrical dielectric waveguide. The physical parameters $\lambda$, $d$, $n_1$, and $n_2$ specify the relationship between the eigenvalues $u$ and $q$ through equation (19). The eigenvalues are determined for each $n$ and $m$ which are solutions of the transcendental eigenvalue equation (17). These values, along with the definition of $\alpha$ in equation (16), are used to evaluate the fields in each medium through equations (14). In medium 1, equations (14) can be used directly using mode constants with subscript 1. Outside the guide, we use equations (14) with the modifications of equation (15).

## *Power Flow through the Dielectric Waveguide Cylinder*

Of more direct interest than the fields themselves may be the flow of electromagnetic energy through the waveguide. The axially propagated power due to the fields is given by the time-averaged axial component of the Poynting vector:

$$
\begin{aligned}
\bar{S}_z &= \frac{1}{2}\mathrm{Re}\left(\bar{E} \times H^*\right)_z \\
&= \frac{1}{4}\mathrm{Re}\left[i\left(E_+H_+^* - E_-H_-^*\right)\right]
\end{aligned}
\tag{20}
$$

Note that the left- and right-circularly polarized components contribute independently to the axial power flow. The power flow inside the waveguide, $P_1$, is given by the integration:

$$
P_1 = \int_0^{2\pi}\int_0^a \bar{S}_z \rho \, d\rho \, d\phi
\tag{21a}
$$

and outside the guide, the power flow, $P_2$, is given by integrating over the space exterior to the cylinder:

$$P_2 = \int_0^{2\pi} \int_a^{\infty} \overline{S}_z \rho d\rho d\phi, \tag{21b}$$

where we are explicitly using $a = d/2$ to specify the radius of the cylinder.

The power flow along the outside of the cylinder occurs as an evanescent surface wave whose amplitude decreases rapidly (exponentially) with increasing radial distance from the cylinder wall. We explicitly compute the radial distribution of the power inside and outside a waveguide (before conducting the integrations of equations 21a and 21b) for the $HE_{11}$ mode (Figure 51) and for the $HE_{21}$ mode (Figure 52).

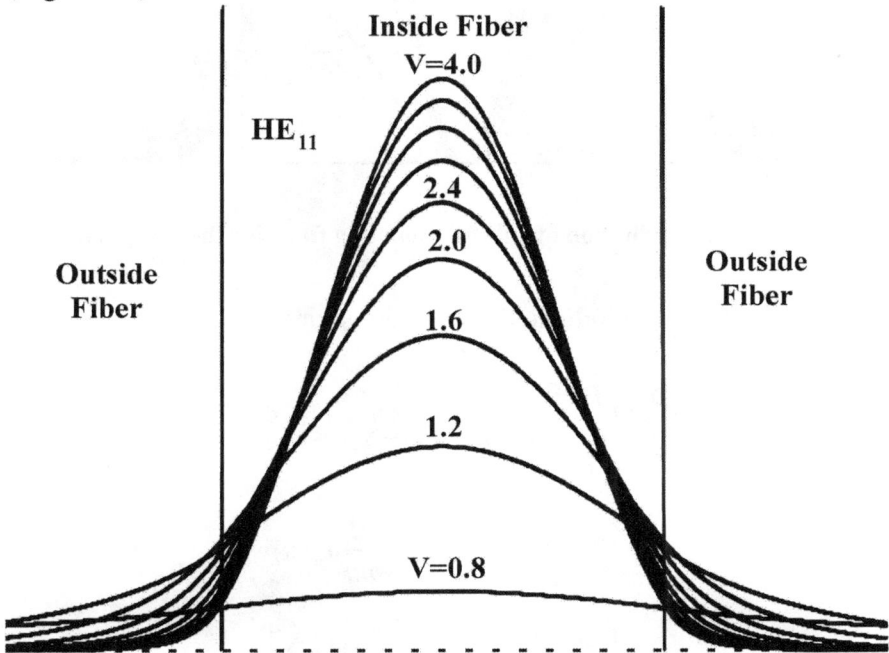

**Figure 51. Radial distribution of light illuminating fiber for the $HE_{11}$ mode**

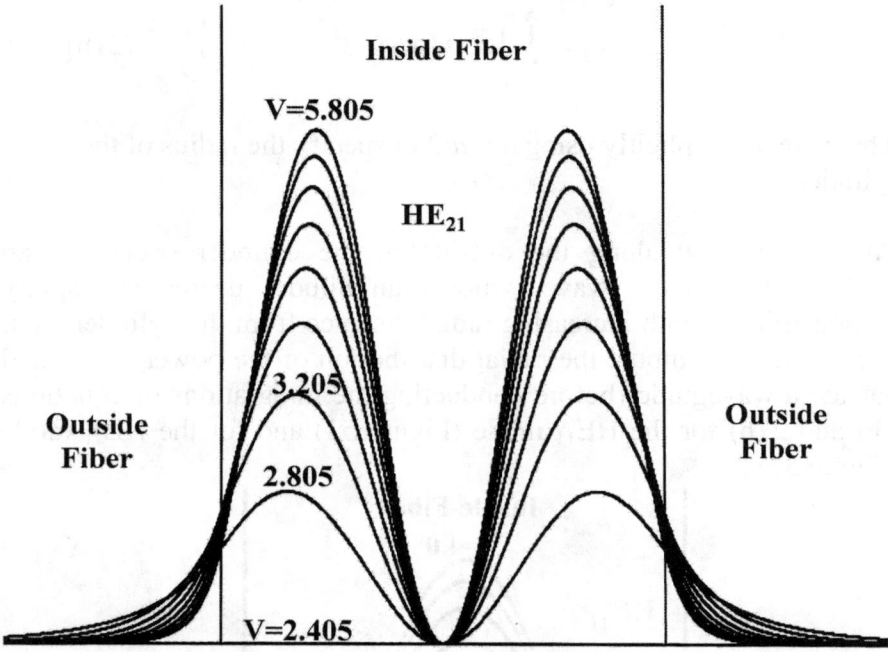

**Figure 52. Radial distribution of light illuminating fiber for the HE$_{21}$ mode**

Now, carrying out the indicated integrations yields the result:

$$P_1 = P_0\left(J_n^+ + \gamma J_n^-\right) \tag{22a}$$

$$P_2 = P_0\left(-K_n^+ - \gamma K_n^-\right) \tag{22b}$$

with

$$P_0 = \left(AA^*\right)\frac{1+\alpha}{\zeta^2}\frac{\pi h d}{4\omega\mu}J_n^2(u) \tag{23}$$

$$J_n^{\pm} = \left(1 \pm \zeta^2\alpha\right)\left\{\left[\frac{J_{n\pm1}^2(u)}{J_n^2(u)}\right] - 2(n \pm 1)\left[\frac{J_{n\pm1}(u)}{uJ_n(u)}\right] + 1\right\} \tag{24}$$

$$K_n^{\pm} = \left(k^2 \pm \zeta^2\alpha\right)\left\{\left[\frac{K_{n\pm1}^2(q)}{K_n^2(q)}\right] - 2(1 \pm n)\left[\frac{K_{n\pm1}(q)}{qK_n(q)}\right] - 1\right\} \tag{25}$$

166

$$k^2 = k_2^2 / k_1^2 = \varepsilon_2 / \varepsilon_1 = 1 - \delta \qquad (26)$$

$$\zeta^2 = h^2 / k_1^2 = 1 - (u/V)^2 \delta \qquad (27)$$

$$h^2 = k_1^2 - \beta_1^2 = \frac{4}{d^2} \left( \frac{V^2}{\delta} - u^2 \right) \qquad (28)$$

$$\gamma = \frac{1 - \alpha}{1 + \alpha} \qquad (29)$$

From equations (21a) and (21b) we see that the polarization properties of the fields (relative contributions of the left- and right-circularly polarized components) is characterized by the ratio $\gamma$ (equation 29) where $\alpha$ is given by equation (16).  Of particular interest is the waveguide modal efficiency specifying that fraction of the total power propagated in the mode which is conducted within the waveguide. This fraction, $\eta$, is given by

$$\eta = \frac{P_1}{P_1 + P_2} = (1 + P_2 / P_1)^{-1} \qquad (30)$$

where $P_1 + P_2$ is the total power in the mode.

Larger values of the waveguide parameter, $V$, which is commonly written as

$$V = \frac{\pi d}{\lambda} \left( n_1^2 - n_2^2 \right)^{1/2}, \qquad (31)$$

correspond to a larger permissible range of eigenvalues, $u$, which will be solutions of equation (17) with progressively larger values of the modal index $n$.  That is, with increasing $V$, more propagation modes are solutions of the waveguide boundary conditions.  The actual power flow distribution in a fiber with large $V$ will depend on which of the permissible modes are actually excited in the waveguide.  This will, in general, be some complicated function of the coupling conditions at the input end of the guide.

167

John A. Medeiros

## The HE₁₁ mode for the Small Waveguide Case

In small waveguides with $V < 2.405$ the theoretical interpretation will be entirely unambiguous. Whatever energy is guided by the fiber will be entirely in the $HE_{11}$ mode. This case is of particular interest to a number of physical situations and we will examine the n=1 case in more detail here. In this case, the eigenvalue equation (17) reduces to the form:

$$\frac{J_0(u)}{uJ_1(u)} = \frac{1}{u^2} + (1-\delta/2)\left(\frac{K_0(q)}{qK_1(q)} + \frac{1}{q^2}\right) - \left[\frac{\delta}{4}\left(\frac{K_0(q)}{qK_1(q)} + \frac{1}{q^2}\right)^2 + \frac{V^2}{u^2q^4}\left(\frac{V^2}{u^2q^4} - \delta\right)\right]^{1/2}, \quad (32)$$

where the eigenvalue $u_{nm}^{HE}$ is the smallest non-zero value satisfying the above equation.

Now the power flow ratio $P_2/P_1$ is given by:

$$\frac{P_2}{P_1} = -\frac{u^2}{q^2}\frac{K_1^+(q) + \gamma K_1^-}{J_1^+(q) + \gamma J_1^-} \qquad (33)$$

with the $J_1^{\pm}$ and $K_1^{\pm}$ functions given by:

$$J_1^+ = \left(1 + \zeta^2\alpha\right)\left(\frac{J_0^2(u)}{J_1^2(u)} + 1 - \frac{4}{u^2}\right) \qquad (34)$$

$$J_1^- = \left(1 - \zeta^2\alpha\right)\left(\frac{J_0^2(u)}{J_1^2(u)} + 1\right) \qquad (35)$$

$$K_1^+ = \left(k^2 + \zeta^2\alpha\right)\left(\frac{K_0^2(q)}{K_1^2(q)} - 1 - \frac{4}{q^2}\right) \qquad (36)$$

168

$$K_1^- = \left(k^2 - \zeta^2 \alpha\right)\left(\frac{K_0^2(q)}{K_1^2(q)} - 1\right), \qquad (37)$$

where $\alpha$ for $n = 1$ is:

$$\alpha = \left(\frac{1}{u^2} + \frac{1}{q^2}\right)\left\{\frac{J_0(u)}{uJ_1(u)} - \frac{K_0(q)}{qK_1(q)} - \frac{1}{u^2} - \frac{1}{q^2}\right\}^{-1}. \qquad (38)$$

The $HE_{11}$ modal efficiency, $\eta = 1/(1+P_2/P_1)$ gives the same (steep) cut off characteristics typical of the higher-order waveguide modes, except where $\eta$ is very small. For $V$ large, $\eta$ approaches unit efficiency (very little power resident outside the cylinder).

For all waveguide modes, as $V$ decreases toward a critical cut off value, the modal efficiency drops to zero (no power flow within the cylinder). This cut off is quite abrupt for all higher order modes; for the $HE_{11}$ mode, while $\eta$ is not exactly zero until $V = 0$, it does get very small for finite values of $V$ below 1.0. For example, at $V = 0.3$, $\eta \approx 10^{-16}$!

## Case of Negligible Refractive Index Difference and Analytic Approximations

It is interesting and informative to examine the limiting case of $\delta \to 0$. In this limit of negligible difference in refractive index between the cylinder and its surround, the eigenvalue equation (32) simplifies considerably to the form:

$$\frac{J_0(u)}{uJ_1(u)} = \frac{K_0(q)}{qK_1(q)}. \qquad (39)$$

This is a useful result since many cases of practical interest involve relatively small dielectric difference, $\delta$ (on the order of 0.1 or less) and the results do not markedly differ from that of this small-$\delta$ limit.

It is also possible to obtain an analytic expression for the eigenvalues in the limit of small $V$ through the use of equation (39). Snyder (1969) noted that in the limit of small $V$ where $u$ approaches $V$ (and thus $q \to 0$) we may reasonably replace $u$ by $V$ in the equation. Using the small argument asymptotic form of the Bessel functions in the right-hand side of the equation in the form

$$\frac{K_0(q)}{qK_0(q)} = \ln(1.123/q),$$

we get

$$q = 1.123 \exp\left(\frac{-J_0(V)}{VJ_0(V)}\right).$$

The eigenvalue $u$ may then be computed through $u^2 = V^2 - q^2$. This analytic approximation can be very accurate for small $V$.

Snyder (1969) also derived an even simpler analytic expression for the asymptotic limit of large $V$. In this case the modal efficiency is accurately represented for $V$ greater than about 1.5 by the expression:

$$\eta = 1 - \frac{5.784}{V^3} e^{-2/V}$$

## *"Back of the Envelope" Polynomial Representation*

For the $HE_{11}$ mode it is possible to do a polynomial approximation through curve fitting in order to get an approximate analytical representation for quick (and very rough) computational purposes. Such a representation, using a sixth order polynomial, for the waveguide efficiency as a function of waveguide parameter is:

$$\eta = 0.0276V^6 - 0.4021V^5 + 2.3356V^4 - 6.8514V^3 + 10.397V^2 - 6.974V + 1.643$$

This representation gives efficiency values (for $HE_{11}$ mode, only) accurate to better than 1% for values of V in the useful range: $0.55 < V < 3.0$.

# CONE SHAPE AND COLOR VISION:
## UNIFICATION OF STRUCTURE AND PERCEPTION

# REFERENCES

Abramov, I., J. Gordon, and H. Chan (1991) "Color appearance in the peripheral retina: effects of stimulus size" *J. Opt. Soc. Am. A* **8**: 404-414

Abramov, I., J. Gordon, and H. Chan (1992) "Color appearance across the retina: effects of a white surround" *J. Opt. Soc. Am. A* **9**: 195-202

Adelson E.H. (1978) "Iconic storage: the role of rods" *Science* **201**: 544-546

Ahnelt, P. K., H. Kolb, and R. Pflug (1987) "Identification of a subtype of cone photoreceptor, likely to be blue sensitive, in the human retina" *J. Comp. Neurol.* **255**: 18-34.

Allen, F. (1926) "The persistence of vision" *Am. J. Physiol. Optics* **7**: 439-457

Alpern, M. (1974) "What is it that confines in a world without color?" *Invest. Ophthal.* **13**: 647-674

Alpern, M. (1986) "The Stiles-Crawford effect of the second kind (SCII): a review" *Perception* **15**: 785-799

Alpern, M, G.B. Lee, F. Maaseidvaag, and S.S. Miller (1971) "Colour vision in blue-cone 'monochromacy'" *J. Physiol.* **212**: 211-233

Alpern, M., G.B. Lee, and B.E. Spivey (1965) "Cone monochromatism", *Arch. Ophthalmol.* **74**: 334

Baker, H.D., and W.A.H. Rushton (1965) "The red sensitive pigment in normal cones" *J. Physiol.* **176**: 56

Balaraman, S. (1962) "Color Vision Research and the Trichromatic Theory: A Historical Review" *Psych. Bulletin* **59**: 434-448

Barer, R. (1957) "Refractometry and Interferometry of living cells" *J. Opt. Soc. Am.* **47**: 545-56

Bartlett, N.R. (1965) "Dark and light adaptation" Ch. 8, Vision and Visual Perception, C.H. Graham (ed.), John Wiley and Sons, Inc., NY

Bidwell, S. (1901), "On the negative afterimages and their relation to certain other visual phenomena" *Proc. Roy. Soc.* **68**: 262 – 285

Biernson, C. (1966), "Evaluation of physiological evidence for trichromatic theory" Cybernetic Problems in Bionics, Oestreicher, H.L. and Moore, D.R., eds., Bionics Symposium, Gordon and Breach,. N.Y. (1968) 407-417

Biernson, C. and A.W. Snyder (1968) "A model of vision employing optical mode patterns for color discrimination" *IEEE Trans.* **SSC-4**: 173-181

Born, N. and E. Wolf (1959) Principles of Optics, Pergamon Press, NY

Borwein, B. (1981) "The retinal receptor: A description" Vertebrate Photoreceptor Optics, J.M. Enoch and F.L. Tobey, eds., Springer-Verlag, NY, 11-81

Borwein, B., D. Borwein, J.A. Medeiros, and J. W. McGowan (1980) "Ultrastructure of monkey foveal photoreceptors with special reference to the structure, shape, size, and spacing of foveal cones" *Am. J. Anat.* **159**: 125-146

Boynton, R.M., W. Schafer and M.A. Neun (1964) "Hue-wavelength relation measured by color naming method for three retinal locations" *Science* **146**: 666-668

Boynton, R.M., H.M. Scheibner (1967) "On the perception of red by 'red-blind' observers" *Acta Chromatica* **1**: 205-220

Brindley, G.S. (1953) "The effects on colour vision of adaptation to very bright lights" *J. Physiol.* **122**: 332-350

Brindley, G.S. (1960) <u>Physiology of the Retina and Visual Pathway,</u> Edward Arnold (Pub.) Ltd., London

Brindley, G.S. and W.A.H. Rushton (1959) "The color of monochromatic light when passed into the human retina from behind" *J. Physiol.* **147**: 204-208

Brown, P.X. and G. Wald (1964) "Visual pigments in single rods and cones of the human retina" *Science* **144**: 45-52

Brücke, E.W. (1843) "Beiträge zur Lehre von der Diffusion tropfbar flüssiger Körper durch poröse Scheidewände" *Ann. Phys. Chem.* **53**: 77-94

Bures, J. and R, Ghosh (1999) "Power density of the evanescent field in the vicinity of a tapered fiber" *J. Opt. Soc. Am. A* **8**: 1992-1996

von Campenhausen, C. and J. Schramme (1995) "100 years of Benham's Top in colour science" *Perception* **24**: 695-717

Choi, S.S., N. Doble, J. Lin, J. Christou, and D.R. Williams (2005) "Effect of wavelength on the *in vivo* images in the human cone mosaic" *J. Opt. Soc. Am. A* **22**: 2598-2605

Clarke, F.J.J. (1960) "A Study of Troxler's effect" *Optica Acta* **7**: 219-236

Cohen, A.I. (1961) "The fine structure of the extrafoveal receptors of the rhesus monkey" *Exp. Eye Res.* **1**: 128-136

Cohen, A.I. (1972) "Rods and Cones" Handbook of Sensory Physiology VII/2, Springer-Verlag, N.Y. 63-110

Cohen, J. and D.A. Gordon (1949) "The Prevost-Fechner-Benham Subjective Colors" *Psych. Bul.* **46**: 97-133

Coltheart M. (1980) "The persistences of vision" *Philos. Trans. R. Soc. Lond. B. Biol. Sci.* **290**: 57-69

Crawford, B.H. (1972) "The Stiles-Crawford Effects and their Significance in Vision" Handbook of Sensory Physiology **VII/4**: 470-483

Crescitelli, F. (1972) "The visual cells and visual pigments of the vertebrate eye", Handbook of Sensory Physiol. **VII/1**: 245-363

Curcio, C.A., K.A. Allen, K.R. Sloan, C.L. Lerea, J.B. Hurley, IB. Klock, and A.H. Milam (1991) "Distribution and morphology of human cone photoreceptors stained with anti-blue opsin", *J. Comp. Neurol.* **312**: 610-624

Curcio, C.A., K.R. Sloan, R.E. Kalina, and A.E. Hendrickson (1990) "Human photoreceptor topography", *J. Comp. Neurol.* **292**: 497-523

Curcio, C.A., K.R. Sloan, Jr., O. Packer, A.E. Hendrickson, and R.E. Kalina (1987) "Distribution of cones in the human and monkey retina: individual variability and radial asymmetry", *Science* **236**: 579-582

Dartnall, H.J.A. (1960) "Visual pigments of colour vision", Mechanisms of Colour Discrimination, Proc. 1958 Paris Symposium, Pergamon Press, N.Y.: 147-177 di Francia, T. G. (1949) "Retinal cones as dielectric antennas" *J. Opt. Soc. Am.* **39**: 324

Ditchburn, R.W. (1956) "Eye Movements and Visual Perception", *Research* **9**: 466-471

Ditchburn, R.W. (1957) "Eye-movements in relation to perception of colour" N.P.L. Symposium on Problems of Colour Vision, H.M. Stationery Office, London, pp.415-427

Ditchburn, R.W. (1961) "Eye-Movements in Relation to Perception of Colour" Visual Problems of Color Symposium, Vol. II, Paper No. 15, Chemical Publishing Co., NT

Ditchburn, R.W. (1973) Eye-Movements and Visual Perception, Clarendon Press, Oxford

Ditchburn, R.W. and D.H. Fender (1955) "The Stabilized Retinal Image" *Opt. Acta.* **2**: 128-133

Dowling, J.E. (1965) "Foveal receptors of the monkey retina: Fine Structure" *Science* **147**: 57-59

Elsner, A.E., S.A. Burns, and R.H. Webb (1993) "Mapping cone photopigment optical density" *J. Opt. Soc. Am. A* **10**: 52-58

Enoch, J. M. (1961a) "Visualization of waveguide modes in retinal receptors" *Am J. Ophthalmol.* **51**: 1107-1118

Enoch, J.M. (1961b) "Nature of transmission of energy in the retinal receptors" *J. Opt. Soc. Am.* **51**: 1122-1126

Enoch, J.M. (1963) "Optical properties of the retinal receptor" *J. Opt. Soc. Am.* **53**: 71-85

Enoch, J.M. (1966) "Retinal microspectrophotometry" *J. Opt. Soc. Am.* **56**: 833-835

Enoch, J.M. (1967) "The Retina as a Fiber Optics Bundle, Appendix B" Fiber Optics, Principles and Applications, N.S. Kapany, ed., Academic Press, NY: 372-396

Enoch, J.M. (1972) "The Two-color threshold technique of Stiles" Handbook of Sensory Physiol **VII/4** 537-565

Enoch, J.M. and L.E. Glismann (1966) "Physical and optical changes in excised retinal tissue" *Invest. Ophthal.* **5**: 208-211

Enoch, J.M. and W.S .Stiles (1961) "The colour change of monochromatic light with retinal angle of incidence" *Optica Acta* **8**: 329-358

Enoch, J.M., J. Scandrett and F.L.J. Tobey (1973) "A study of the effects of bleaching on the width and index of refraction of frog rod outer segments" *Vision Research* **13**: 171-183

Falls, H.F., R. Wolter and M. Alpern (1965) "Typical total monochromacy - A histological and psychophysical study" *Arch. Ophthal.* **74**, 610-616

Fechner, G.T. (1838) "Ueber eine Scheibe zur Erzeugung subjectiver Farben" Annalen der Physik und Chemie, J.C. Poggendorf, (ed.), pp. 227-232, Verlag von Johann Ambrosius Barth, Leipzif

Feeney, L. (1973a,b) "The Interphoreceptor Space Postnatal Ontogeny in mice and rats: II. Histochemistry of the Matrix" *Developmental Biology* **32**: 101-114; 115-128

Festinger, L., M.R. Allyn, and C.W. White (1971) "The perception of color with achromatic stimulation" *Vision Research* **11**: 591-612

Fine, B.S. and L.E. Zimmerman (1963) "Observations on the rod and cone layer of the human retina: A Light and electron microscopic study" *Invest. Ophthal.* **2**: 446-459

Fischer, W. and R. Rohler (1974) "The absorption of light in an idealized photoreceptor on the basis of waveguide theory II: The semi-infinite cylinder" *Vision Research* **14**: 1115-1125

Frölich, F.W. (1921) "Untersuchungen über periodischen Nachbilder" *Z. f. Sinnes physiol.* **52**, 60 – 88

Frölich, F.W.(1922) "Über der Hell – und Dunkeladaptation auf den Verlauf der periodischen Nachbilder" *Z. f. Sinnes physiol.* **53**, 88–104

Graham, C.H. and Y. Hsia (1958) "Color defect and color theory" *Science* **127**; 675-682

Gibson, K.S. and E.P.T. Tyndall (1923) "Visibility of radiant energy" *Papers of the Bureau of Standards* **19** No. 475, 131

Glickstein, M. and Heath, C.G. (1975) "Receptors in the monochromat eye" *Vision Research* **15**: 633-636

Gray-Keller, M., W. Denk, B. Shraiman, and P.B. Detwiler (1999) "Longitudinal spread of second messenger signals in isolated rod outer segments of lizards" *J. Physiol.* **519**: 679-692

von Greef, R. (1900) "Mikroskopische anatomie des secherven unde netzhaut" Graefe-Saemisch Handbuch d. ges. Augenheilk, 2. Avfl. 1. Teil. bd. 1, Kap. U.

Hagins, W.A., R.D. Penn, and S. Yoshikami (1970) "Dark current and photocurrent in retinal rods" *Biophys. J.* **10**: 380-412

Hall, M.O. and J. Heller (1969) "Mucopolysaccharides of the retina" *UCLA Forum Med. Sci.* **8**: 211-224

Hamaker, H.G. (1899). "Ueber Nachbilder nach momentanner

Helligkeit" *Zsch. F. Psychol. u. Physiol. D. Sennesorg* **21**: 1-44

Hamilton, S.E. and J.B. Hurley (1990) "A phosphodiesterase inhibitor specific to a subset of bovine retinal cones" J. Biol. Chem. 265: 11259-11264

Hárosi, F.I. and E.F. MacNichol, Jr. (1974) "Visual pigments of goldfish cones, spectral properties and dichroism" *J. Gen. Physiol.* **63**: 279-304

Harrick, N.J., (1967). Internal Reflection Spectroscopy, Interscience Publishers, div. of John Wiley & Sons, NY

Harrison, R., D. Hoefnagel and J.N. Hayward (1960) "Congenital Total Color Blindness" *Arch. Ophthal.* **64**: 685-692

von Helmholtz, H. (1924) Treatise on Physiological Optics, Vols. I-III, James P.C. Southhall (Translator), Optical Society of America

Hemila, S. and T. Reuter (1981) "Longitudinal spread of adaptation in the rods of the frog's retina" *J. Physiol.* **310**: 501-528

Hendrickson, A. E. and C. Youdelis (1984) "The morphological development of the human fovea" *Ophthalmol.* **91**: 603-612.

Hofer, H. and D.R. Williams (2002) "The eye's mechanism for autocalibration" *Optics & Photonics News*, January 2002: 34-39

Hofer, H., B. Singer, and D.R. Williams (2005) "Different sensations from cones with the same photopigment" *Journal of Vision* **5**: 444-454

Holcman, D. and J.I. Korenbrot (2004) "Longitudinal diffusion in retinal rod and cone outer segment cytoplasm: The consequence of cell structure" *Biophys. J.* **86**: 2566-2582

Hsia, Y. and C.H. Graham (1957) "Spectral luminosity curves for protanopic, deuteranopic and normal subjects" Proc. Nat. Acad. Sci. 43: 1011-1019

Hurvich, L.M. and D. Jameson (1957) "An opponent-process theory of color vision," *Psychol. Rev.* **64**: 384-402

Ives, H.E. (1918) "The resolution of mixed colors by differential visual diffusivity" *Phil. Mag.* **35**: 413-421

Jameson K.A., S.M. Highnote, and L.M. Wasserman (2001) "Richer color experience in observers with multiple photopigment opsin genes" *Psychon. Bull. Rev.* **8**: 244-261

Jameson, D. and L.M. Hurvich (1978) "Dichromatic color language: 'reds' and 'greens' don't look alike but their colors do" *Sensory Processes* **2**: 146-155

Judd, D.B.(1927) "A quantitative investigation of the Purkinje after-image" *Amer. J. Psychol.* **38**: 507–533

Kalmus, H. (1965) Diagnosis and Genetics of Defective Colour Vision, Pergamon Press, Oxford

Kapany, N.S. (1967) Fiber Optics, Principles and Applications, Academic Press, NY

Kapany, N.S. and Burke, J.J. (1972) Optical Waveguides, Academic Press, NY

Karwoski, T. and M.N. Crook (1937) "Studies in the peripheral retina I: The Purkinje after-image" *J. Gen. Psychol.* **16**: 259–264

Karwoski, T. F. and H. Warrener (1942) "Studies in the peripheral retina II: The Purkinje after-image on the near foveal area of the retina" *J. Gen. Psychol.* **26**: 129–151

Keele, S.W. and W. G. Chose (1967) "Short-term visual storage" *Perception and Psychophysics* **2**: 383–386

Kinney, J.A., and C.L. McKay (1974) "Test of color-defective vision using the visual evoked response" *J. Opt. Soc. Am.* **64**: 1244-1150

Kraft, J. M., and J. S. Werner (1999) "Aging and the saturation of colors. 1. Colorimetric purity discrimination" *J. Opt. Soc. Am. A* **16**: 223-230

Kriegman, D.H. and I. Biederman (1980) "How many letters in Bidwell's ghost? An investigation of the upper limits of full report from a brief visual stimulus" *Perception and Psychophysics* **28**: 82–84

von Kries, J. A. (1905) "Die Gesichtsempfindungen" In Nagel, W. (Ed.), Handbuch der Physiologie des Menschen. Vol. 3, Physiologie der Sinne. Fredrich Vieweg und Sohn, Braunschweig, 109-282

Kurtenbach, A., S. Meierkord, and J. Kremers (1999) "Spectral sensitivities in dichromats and trichromats at mesopic retinal illuminances" *J. Opt. Soc. Am. A* **16**: 1541-1548

Lamb, T.D., P.A. McNaughton, and K.W. Yau (1981) "Spatial spread of activation and background desensitization in toad rod outer segments" *J. Physiol.* **319**: 463-496

Larsen, H (1921) "Demonstratin mikroskopischer Praparate von einem monochromatischen Auge" *Klin. Monstabl. Augenheilk* **67**: 301

Liebman, P.A. (1972) "Microspectrophotometry of photoreceptors" Handbook of Sensory Physiology VII/1, pp 482-528, Springer

Verlag, N.Y.

Long, G.M. (1980) "Iconic memory: a review and critique of the study of short-term visual storage" *Psychological Bulletin* **88**: 785–820

Marcuse, D. (1970) "Radiation losses of the dominant mode in round dielectric waveguides" *Bell Syst. Tech. J.* **49**: 1665-1693

Marcuse, D. (1974) Theory of Dielectric Optical Waveguides, Academic Press, N.Y.

Marks, W.B., Dobelle, W.H., MacNichol, E.F., Jr. (1964) "Visual Pigments of Single Primate Cones" *Science* **143**: 1181-1183

Matthews, G. (1986) "Spread of the light response along the rod outer segment: an estimate from patch-clamp recordings" *Vision Research* **26**: 535-541

McDougall, W. (1904a) "The sensations excited by a single momentary stimulation of the eye" *Brit. J. Psychol.* **1:** 78–113

McDougall, W. (1904b) "The variation of the intensity of visual sensation with the duration of the stimulus" *Brit. J. Psychol.* **1**: 151–189

McDougall, W. (1904c) "The illusion of the 'fluttering heart' and the visual functions of the rods of the retina" *Brit. J. Psychol.* **1**: 428–434

McMahon, C., J. Neitz, and M. Neitz (2004) "Evaluating the human X-chromosome pigment gene promoter sequences as predictors of L:M cone ratio variation" Journal of Vision 25: 203-208

Medeiros, J.A. (1979) "Optical transmission in photoreceptors: implications for SC II Effect" *J. Opt. Soc. Am.* **69**: 1486

Medeiros, J.A., B. Borwein and J. W. McGowan (1977) "Spectroscopic characteristics of small dielectric cones" *J. Opt. Soc. Am* **67**: 1372

Medeiros, J.A., G.C. Caudle, and N.E. Schildt (1982) "Novel visual effect elicited by a moving slit of light" *J. Opt. Soc. Am.* **72**: 1741

Miller, W.H. and Snyder, A.W. (1972) "Optical function of human peripheral cones" *Vision Research* **13**: 2185-2194

Missotten, L. (1974) "Estimation of the ratio of cones to neurons in the fovea of the human retina" *Invest. Ophthal.* **13**: 1045-1049

Mollon, J.D. and G Estévez (1988) "Tyndall's paradox of hue discrimination" *J. Opt. Soc. Am. A* **5**: 151-159

Moreland, J.D. (1972) "Peripheral Colour Vision" <u>Handbook of Sensory Physiology</u> VII/4, pp 517-533

Moreland, J.D. and Cruz, A. (1959) "Color perception with the peripheral Retina" *Optica Acta* **6**: 117-151

Müller, B., L. Peichl, W.J.D. Grip, I. Grey, and H.W. Korf (1989) "Opsin- and S-antigen-like immunoreactions in photoreceptors of the tree shrew retina" *Invest. Ophthal. Vis. Sci.* **20**: 530-535

Myers, O. (1962) "Spectral sensitivity of visual receptor cells" , *Nature (Lond.)* **193**: 449-451

Nathans, J. (1999) "The evolution and physiology of human color vision: Insights from molecular genetic studies of visual pigments" *Neuron* **24**: 299-312

Nathans, J., D. Thomas, and D.S. Hogness (1986) "Molecular genetics of human color vision: The genes encoding blue, green and red pigments" *Science* **232**: 193-202

Nathans, J., T.P. Piantanida, R.L. Eddy, T.B. Shows, and D.S. Hogness (1986) "Molecular genetics of inherited variation in human color vision" *Science* **232**: 203-210

Neitz, M. and J. Neitz (1995) "Numbers and ratios of visual pigment genes for normal red-green color vision" *Science* **267**: 1013-1016

Neitz, M. and J. Neitz (2000) "Molecular genetics of color vision and color vision defects" *Arch. Ophthalmol.* **118**: 691-700

Neitz, M, J. Neitz J, and A. Grishok (1995) "Polymorphism in the number of genes encoding long-wavelength-sensitive cone pigments among males with normal color vision" Vision Research 35: 2395-2407

Neitz, J., M. Neitz, J.C. He, and S.K. Shevell (1999) "Trichromatic color vision with only two spectrally distinct photopigments" *Nature Neuroscience* **2**: 884-888

Neitz, M., J. Neitz, and G.H. Jacobs (1989). "Analysis of fusion gene and encoded photopigment of colour-blind humans" Nature 342: 679-682.

Neitz, M, J. Neitz, and G.H. Jacobs (1993) "More than three different cone pigments among people with normal color vision" *Vision Research* **33**: 117-122

Nijhawan, R (1997) "Visual decomposition of colour through motion extrapolation" *Nature* **386**: 66-69

Nork, T.M., S.A. McCormick, G.M. Chao, and J.V. Odom (1990) "Distribution of carbonic anhydrase among human photoreceptors" *Invest. Ophthal. Vis. Sci.* **31**: 1451-1458

O'Brien, B. (1946) "A theory of the Stiles and Crawford Effect" *J.*

*Opt. Soc. Am.* **36**: 506-509

O'Brien, B. (1951) "Vision and resolution in the central retina" *J. Opt. Soc. Am.* **41**: 882-894

Østerberg, G. (1935) "Topography of the layer of rods and cones in the human retina" *Acta. Ophthal.* kbh., Supp. 6

Pankove, J.I. (1971) Optical Processes in Semiconductors, Prentice-Hall, Englewood Cliffs, N.J.

Pitt, F.G.H. (1935) "Characteristics of dichromatic vision, with an appendix on anomalous trichromatic vision" *Great Britain Med. Research Council, Special Rpt. Ser.* **200**

Polizzotto, L. and Peura, R.A., (1975) "A mathematical approach to explain subjective color perception" *Vision Research* **15**: 613-616

Polyak, S.L., (1941) The Retina, Univ. of Chicago Press, Chicago, Ill.

Priest, I. G., and F. G. Brickwedde (1938) "Minimum perceptible colorimetric purity as a function of dominant wavelength" *J. Opt. Soc. Am.* **28**: 133-139

Purdy, D.M. (1931) "On the saturations and chromatic thresholds of the spectral colours" *Br. J. Psychol.* **21**: 283-313

Ratliff, F. and L. A. Riggs (1950) "Involuntary Motions of the Eye During Monocular Fixation" *J. Exp. Psychol.* **40**: 687-701

Raymond, L.F. (1971) "Color blindness" *Ann Allergy.* **29: 2**14-216

Raymond, L.F. (1972) "Color-blindness-deficiency disease, auto-immune, or sex-linked inherited deficiency disease" *Eye Ear Nose Throat Mon.* **51: 1**46-1477.

Raymond, L.F. (1975) "Physiology of color vision and the pathological changes in reversible color blindness, a deficiency disease of the retina" *Ann Ophthalmol.* **7**: 532-534

Riggs, L. (1967) "Electrical evidence on the trichromatic theory" *Invest. Ophthal.* **6**: 6-17

Riggs, L. A. and F. Ratliff (1951) "Visual Acuity and the Normal Tremor of the Eyes" *Science* **114**: 17-18

Riggs, L.A., F. Ratliff, J.C. Cornsweet, and T.N. Cornsweet (1953) "The Disappearance of Steadily Fixated Visual Test Objects" *J. Opt. Soc. Am.* **43**: 495-501

Roorda, A., and D.R. Williams (1999) "The arrangement of the three cone classes in the living human eye" *Nature* **397**: 520-522

Rowe, M.P., J.M. Corless, N. Engheta, and E.N. Pugh, Jr. (1996) "Scanning Interferometry of sunfish cones. I. Longitudinal variation in single-cone refractive index" *J. Opt. Soc. Am. A* **13**: 2141-2150

Sakitt, B. (1976) "Iconic memory" *Psychol. Rev.* **83**: 257–276

Sanford (1903) Experimental Psychology, Boston, p. 318

Scheibner, H.M., R.M. Boynton (1968) "Residual red-green discrimination in dichromats" *J. Opt. Soc. Am.* **58**: 1151-1158

Schnapf, J.L., B.J. Nunn, M. Mester, and D.A. Baylor (1990) "Visual transduction in cones of the monkey Macaca fasicularis" J. Physiol. (Lond.) 427: 681-713

Schroeder, A.C. (1960) "Theory on the Receptor Mechanism in Color Vision" *J. Opt. Soc. Am.* **50**: 945-949

Sharpe, L.T., A. Stockman, H. Jägle, and J. Nathans (1999) "Opsin genes, cone photopigments, color vision, and color blindness" Color Vision: From Genes to Perception, Cambridge University Press, Cambridge, UK. 3-51

Sheppard, J.J., Jr. (1968) Human Color Perception: A Critical Study of the Experimental Foundation, American Elsevier Publ. Co., New York

Sidman, R. (1957) "The structure and concentration of solids in photoreceptor cells studied by refractometry and interference microscopy" *J. Biophys. Biochem. Cytol.* **3**: 15-30

Snitzer, E. (1961) "Cylindrical dielectric waveguide modes" *J. Opt. Soc. Am.* **51**: 491-498

Snyder, A.W. (1970) "Coupling of modes on a tapered dielectric cylinder" *IEEE Trans.* **MTT-18**: 383-392

Snyder, A.W. (1971) "Mode Propagation in a nonuniform cylindrical medium" *IEEE Trans.* **MTT-19**: 402-403

Snyder, A.W. and J.D. Love (1983) Optical Waveguide Theory, Chapman and Hall, London

Snyder, A.W. and C. Pask (1973) "Waveguide modes and light absorption in photoreceptors" *Vision Research* **13**: 2605-2608

Snyder, AW. and P. Richmond (1972) "Effects of anomalous dispersion on visual photoreceptors" *J. Opt. Soc. Am.* **62**: 1278-1285

Sperling, G. (1960) "The information available in brief visual presentations" *Psychol. Monogr.* **74**: 1

Stavenga, D.G. and H.H. van Barneveld, (1975) "On dispersion in visual photoreceptors" *Vision Research* **15**: 1091-1095

Stewart, G.N. (1924) "Color Phenomena Caused by Intermittent Stimulation with White Light" *Am. J. Physiol.* **69**: 337-353

Stiles, W.S. (1937) "The luminous efficiency of monochromatic rays entering the eye pupil at different points and a new colour effect" *Proc. Roy. Soc. B.* **123**: 90-118

Stiles, W.S. (1959) "Color Vision: the approach through increment threshold sensitivity" *Proc. Nat. Acad. Sci.* **45**: 100

Stiles, W.S. and Crawford, B.H. (1933) "The luminous efficiency of rays entering the eye pupil at different points" *Proc. Roy. Soc. B.* **112**: 428-450

Tyndall, E.T.P. (1933) "Chromatic sensibility to wave-length differences as a function of purity" *J. Opt. Soc. Am.* **23**: 15-22

Vilter, V. (1949) "Recherches biometriques sur l'organisation Synaptique de la retine humaine" *Comp. Rend. Soc. Biol.* **143**: 830

Wachtler, T., U. Dohrmann, and R. Hertel (2004) "Modeling color percepts of dichromats" *Vision Research* **44**: 2843-2855

Wald, G. (1949) "The Photochemistry of Vision" *Doc. Ophthalmol.* **3**: 94

Wald, G. (1967) "Blue-blindness in the normal fovea" *J. Opt. Soc. Am.* **57**: 1289

Walraven, P.L. and M.A. Bouman (1960) "Relation between directional sensitivity and spectral response curves in human cone vision" *J. Opt. Soc. Am.* **50**: 780-784

Weale, R.A. (1953) "Colour vision in the peripheral retina" *Brit. Med.*

*Bull.* **9**: 55-60

Weingarten, F.S. (1972) "Wavelength effect on visual latency" *Science* **176**: 692-694

Wijngaard, W., MA. Bouman and F. Budding (1974) "The Stiles-Crawford colour change" *Vision Research* **14**: 951-957

Williams, D.R. (1988) "Topography of the foveal cone mosaic in the living human eye" *Vision Research* **28**: 433-454

Williams, D.R. (1990) "The invisible cone mosaic" <u>Advances in Photoreception: Proceedings of a Symposium on the Frontiers of Visual Science</u>, pp. 135-148, Washington DC: National Academy Press

Willmer, E.N. and W.D. Wright (1945) "Colour sensitivity of the fovea centralis" *Nature (Lond)* **156**: 119-121

Wooten, B.R., and Wald, G., (1973) "Color-vision mechanisms in the peripheral retinas of normal and dichromatic observers" *J. Gen. Physiol.* **61**: 125-145

Wright, W.D. (1952) "The characteristics of tritanopia" *J. Opt. Soc. Am.* **42**: 509-521

Wright, W.D. (1971) "Small-field tritanopia: A re-assessment" <u>Visual Science, Proc. 1968 Int'l. Symp.</u>, J.R. Pierce and J.R. Levene, eds., Md. U. Press: 152-163

Wright, W.D. and Nelson, J.H. (1936) "The relation between the apparent intensity of a beam of light and the angle at which the beam strikes the retina" *Proc. Phys. Soc. Lond.* **48**: 401-405

Wright, W.D. and Pitt, F.H.G. (1934) "Hue discrimination in normal colour vision" *Proc. Phys. Soc. Lond.* **46**: 459-473

Wright, W.D. and Pitt, F.H.G. (1935) "The colour-vision characteristics of two trichromats" *Proc. Phys. Soc. Lond.* **47**: 205-217

Wright, W.D. and Pitt, F.H.G. (1937) "The saturation discrimination of two trichromats," *Proc. Phys. Soc. Lond.* **49**: 329-331

Xiao, M. and A. Hendrickson (2000) "Spatial and temporal expression of short, log/medium. Or both opsins in human fetal cones" *J. Comp. Neurol.* **425**: 545-559

Yamada, E. (1969) "Some structural features of the fovea centralis in the human retina" *Arch. Ophthal.* **82**: 151-159

Young, Thomas (1802) "On the theory of light and colours" *Philos. trans. Royal Soc., Lond.* **921**: 12

Young, R.S.L. and Alpern, M. (1976) "Heterochromatic brightness matching by pupillography and by flicker photometry" *J. Opt. Soc. Am.* **66**: 1103

Zrenner, E. (1983) <u>Neurophysiological Aspects of Color Vision in Primates,</u> Springer-Verlag, Berlin

Wolken and Pfüt [1967], The colours of...
the... there: two incurred..., Proc. Roy. Soc., Lond. B 166...

Wolf... Marc, Pfütüü... [1967]... Reconstruction..., Immanator...
... ... Thomas... 2004...

Y... ... ... ...and Josephson [1967]... ...tella and Josephson...
... ... ... short ... ... ... quantum... human retina,
... ... ... Nature 263: 573-575.

Y...n ... H. [1972]. Some structure... ... ... fovea cone cells in
the... ... ... J. Physiol. 221: 15-29.

Y...ng Thomas [1802]. On the ... of light and colours, Phil.
Trans. Roy. Soc. ... 92-48.

Y... ... ... and Marc, M. ... ... ... ... mesopic brightness...
... ... pupil... ... ... flicker photometry... Opt...
... ... 36: ... ...

Z... ... ... ... ... ... Studies and essays on Color Vision...
... ... ... by Inger Jelting Barth... 1999.

192

# CONE SHAPE AND COLOR VISION:
## UNIFICATION OF STRUCTURE AND PERCEPTION

# Index

*John A. Medeiros*

*John A. Medeiros*

www.ingramcontent.com/pod-product-compliance
Lightning Source LLC
Chambersburg PA
CBHW071642280326
41928CB00068B/2206